T0335729

Atom Projects

Events and People

Atom Projects

Events and People

Boris Ioffe

ITEP, Moscow, Russia

Translated from Russian by Wladimir von Schlippe

 **World Scientific**

NEW JERSEY · LONDON · SINGAPORE · BEIJING · SHANGHAI · HONG KONG · TAIPEI · CHENNAI · TOKYO

Published by

World Scientific Publishing Co. Pte. Ltd.

5 Toh Tuck Link, Singapore 596224

USA office: 27 Warren Street, Suite 401-402, Hackensack, NJ 07601

UK office: 57 Shelton Street, Covent Garden, London WC2H 9HE

British Library Cataloguing-in-Publication Data
A catalogue record for this book is available from the British Library.

ATOM PROJECTS
Events and People

ISBN 978-981-3145-93-1
ISBN 978-981-3145-94-8 (pbk)

Printed in Singapore

Content

———————————— • ————————————

Introduction

•

Practically all outstanding physicists (as well as chemists and engineers) took part in the Anglo-American atom project, leaving whatever purely scientific work in which they were engaged at the time. They dedicated all their time and all their energy to the atom project, they produced new ideas, they worked with full dedication. And this brought success: suffice it to say that the atom project in Britain and the USA began in 1940, and the first nuclear reactor was started up as early as in 1942. It was understood that plutonium, element $^{249}_{94}X$, could be used as the explosive for the bomb, therefore huge factories were built for the production of plutonium and for the separation of uranium isotopes 235 and 238. At the time of the bombing of Hiroshima and Nagasaki (in 1945), the United States had another 10 nuclear bombs in their arsenal.

There is little discussion in Russian literature of the reasons that inspired the dedication of British and American scientists to their work on the atom project. Neither is there much discussion of the reasons that made so many American scientists leave the atom project once the war against Germany and Japan had ended (the Oppenheimer case).

The intensity of the work done by British and American scientists was so great because of their fear that Hitler's Germany would be the first to produce an atom bomb and that the bomb would be used as a weapon in war. This would have resulted in a worldwide victory for Fascism and the end of democracy.

If Germany won, those scientists who were Jewish could expect total annihilation of all Jews. (The policy of annihilation had indeed been decided by the leaders of Germany in 1943 at the Wannsee conference. It was actively applied in territories under German occupation.)

The political leaders of Britain and the USA clearly understood the danger of Germany building an atomic weapon. This is why Roosevelt, immediately after he received Einstein's letter, ordered the funding of the atom project, created its administrative structure, and later continued to fund the project as it developed.

The Soviet atom project did not have the same motivation. Real action was taken only after Truman informed Stalin at the Potsdam Conference in summer 1945, once Germany had been defeated, that the American atom bomb had already been produced. In September 1945 a Special Committee headed by Beria was created. It included leading members of the Party and Government (and also two scientists — Kurchatov and Kapitza); furthermore the First Primary Directorate (FMD, in Russian PGU of the Sovnarkom — cabinet of Ministers) was formed, headed by Vannikov. Its task was to produce atom bombs. It is usually said that the motivation for the production of the Soviet atom bomb was: since "they" had been able to produce it, then we should also do it. (This is put in the coded name of the Soviet bomb: RDS — Russia will Do it herSelf). Such a view does not stand up to criticism — there was no threat to the USSR.

The USSR must have had a different and strong motivation for devoting its main effort to the creation of the atom bomb in a country that had just survived its greatest ever war, whose population was hungry, badly clothed, and lacked proper housing (many towns in the European part of Russia had been almost completely destroyed).

In my previous book "Without Retouching" (Fazis, 2004) I wrote about some people who had taken part in the Soviet atom project (not all of them: only about the ones whom I had known personally), and described their human characteristics. Their participation in the atom project was not mentioned there. In the present book I describe their participation in the Soviet atom project and (briefly) assess their scientific achievements. My main aim, however, is to show them as living people with their merits and weaknesses, to describe their attitudes towards the atom project and the way in which these changed with time. The main sources which I have used are listed in the complete bibliography, entries [1–9].

This book is meant for the general reader who has no deep knowledge of Physics. Some formulæ are included occasionally, to elucidate a point for a reader familiar with Physics. They may be ignored by the general reader without loss to his understanding of the main content of the book.

B. Ioffe
January 2017

Acknowledgments

I want to thank Professor Wladimir von Schlippe who took the burden of translation of my manuscript from Russian to English. I am very grateful to Elliot Leader for polishing the English text.

Events

2.1. Who Initiated the Second World War? [1]

I will begin with a description of the conditions in the world before
World War II since, as I have said in the Introduction, the Anglo-
American atom project was launched because the leaders of Britain
and the USA feared that Hitler would get an atom bomb.

The Second World War was initiated by Hitler and Stalin — they
wanted to dominate the world or at least half of it. Their declarations were setting out lofty aims. Hitler was going to create a
thousand-year Reich, giving the German nation living space in which
the Germans (Aryans) would be the commanding race, and all the
other races would labour for them and serve them. He particularly
hated the Jews for their sceptical attitude and their lack of obe-
dience to authority, which had its roots in the ancient prophets. In
1935 the Nürnberg laws were proclaimed, according to which Jews
were forbidden to occupy a range of professions. In particular, they
were forbidden to engage in the sciences and to teach in universities.
(Hitler followed the example of the Roman emperor Hadrian, who
barred Jews from engaging in religious sciences after the suppression
of the Bar-Kochba rebellion.) All this caused a mass emigration from
Germany of Jews, and not only of Jews but of all those who were opposed to the Hitler regime. (The Hungarian Jews Leó Szilárd, Eugene
Wigner, Edward Teller and John von Neumann, who had been

working in Germany, left for Britain and the USA and contributed greatly to the atom project — this was a sort of a present that
Hitler gave to his enemies [13].) Such people — Jews and opponents
of the Hitler regime — were put in concentration camps which had
been set up for this purpose. One had to greet each other exclaiming
"Heil Hitler!" and raising one's right arm. Some people who did not
support Hitler sought to avoid this ritual. The Nobel prize winner
Laue, who was not in favour of the Nazi regime, when leaving home
always carried in one hand a briefcase and something else in his other
hand, so both hands were busy.

On the 9th of November 1938 all over Germany and Austria syna-
gogues, enterprises and shops belonging to Jews were destroyed. According to some sources 91 Jews were killed, according to others
many more; many Jews were put in concentration camps (the so-
called "Kristallnacht"). This was the first step towards the "final
solution of the Jewish question", in other words, to the extermina-
tion of the Jews.

Hitler persecuted not only Jews but also those supporters of his
who did not recognize his authority (Röhm, Schleicher and others —
in the "Night of the Long Knives" most of them were murdered).
Anti-Fascists were put in concentration camps and frequently killed.
The kidnapping of anti-Fascists was practised in other countries:
they were forcibly taken to Germany where a tragic fate was awaiting
them.

Stalin wanted to build a worldwide Communist society. Such
"lofty" plans could be realised only by way of a world war.

Hitler came to power in 1933 actually thanks to Stalin. In the
Reichstag elections of November 1932, the National Socialists got
11.7 million votes, the Social Democrats 7.2 million, and the Communists 6 million. If the Social Democrats and the Communists had
presented their candidates for the election jointly, Germany would
have got a democratic government. But Stalin used the Comintern
to give a strong directive to the Communists not to unite with the
Social Democrats. As a result Hitler became the Reichs-Chancellor.
It was exactly with this aim in mind that Stalin had supported Hitler: with a democratic German government there would have been

no world war in Europe, and Stalin's aim would have been frustrated.

For Germany, the first obstacle was the Treaty of Versailles which limited the size of the German army and the assortment of its weapons (tanks were not allowed); a demilitarised zone was established 50 kilometers deep along the right bank of the Rhine. The Saar region was put under the administration of the League of Nations, and 15 years later a referendum was to be held to decide whether it was to become part of Germany or part of France. Various territories of the former German Empire were granted to other countries — France, Poland, Czechoslovakia, Denmark, and Belgium. Hitler's first step in his preparation for a new war was the remilitarization of the Rhine region — German troops were moved into that region. Britain and France confined themselves to diplomatic representations, the US assumed an isolationist policy, and Hitler understood that he could make the next moves.

Stalin thought that preparations for a new war 15 to 20 years after the Civil War would give rise to protests by intellectuals and by the more or less prosperous peasantry, which had just about begun to revive. These peasants (kulaks) were deported to the North and to Siberia, and collective farms (kolkhozes) were established (it is easier to handle kolkhozes than individual peasants). This campaign involved millions of people and caused wide-spread famine — after all, the prosperous peasants were the main producers of bread. Rationing of bread and other produce was introduced in towns.

I have seen this myself. In Moscow in 1933 (I was 7 years old), we were issued grey and sticky bread purchased with ration-cards. But it so happened that I could go with my mother for nine months to Poland. To get permission to travel abroad was very difficult, almost impossible, but we were lucky: an uncle of mine had an acquaintance — A.Kh. Artuzov, who was at that time the Head of the Foreign Department of the OGPU (NKVD). Before the revolution, my uncle, Efim Lvovich Zukerman, had been a barrister in St. Petersburg, and had had an office and a flat there. (Vyshinsky, the Attorney General in 1937 and one of the principal agents of the Great Terror, was an assistant to a barrister. My uncle knew him and spoke of him with great contempt.) My uncle sympathized with the revolutionaries and used to give Artuzov money for their needs. In the Revolution his office and flat were taken away from him, he moved to Moscow and worked as a legal adviser for various Soviet institutions. He had no more contact with Artuzov, but when he approached Artuzov with the request for permission for us to travel to Poland, Artuzov did remember their acquaintance before the revolution and we got our permission. Moreover, we had, of course, no Polish money. But at the Polish border station somebody came up to us and handed my mother some Polish money. In 1937, Artuzov was shot.

Once we had boarded the Polish train, I saw a man in a white apron walking along the gangway with a tray holding cups of coffee and white rolls. I asked my mother to get me some of both. The taste of coffee and rolls is one of the strongest impressions of my childhood.

Then the trials of intellectuals began: the Shakhty trial, the Industrial Party trial, and then the trials of 1935–37. It was not important for Stalin whom to arrest, whom to declare to be an "enemy of the people": what was important was to weaken the intellectuals as a class and to sow fear among the people.

A document has survived in which the Party Secretary of the Kirov District requested permission to increase the quota of the "first category" (i.e. the number of people to be shot) by 300. Stalin crossed out the 300, replaced it by 500 and his signature: I. Stalin. To Stalin it was of no importance who these people were, which among them was guilty and of what. What was important was their number: to spread fear among the people — the fear of a knock at the door at night. (As a rule, people were arrested at night.)

At the same time the deification of the personality of Stalin proceeded.

Large-scale repressions took place among the commanding officers of the Red Army and Navy. About 40 thousand people were repressed: 3 out of 5 Marshals perished, both first-rank Army Commissars, 3 out of 5 first-rank Army Commanders were shot, all of the 12 second-rank Army Commanders, 60 out of the 67 Corps Commanders fell victim to repression, and so on. Repression struck even commanders of regiments and battalions. Even after the war had already started, in October 1941, 16 people were shot, including Yakov Smushkevich, twice Hero of the Soviet Union, Head of the Soviet Air Force (he had fought in Spain). These were replaced by people who were poorly qualified for the job but who were terrified of Stalin and of the Special Services. It is impossible to find an explanation for what Stalin did. He must have realised that he was actually beheading the Red Army. It is inconceivable that he could have genuinely thought that the army was preparing a coup, even less a broadly based one. There is a version that the Gestapo leaked a forged document about Tukhachevsky's treachery to Beneš, the President of Czechoslovakia, who believed it and passed it on to Stalin. But even if one could accept this version, it is inconceivable that as many as 40 thousand people were involved in this conspiracy, and that it had not been disclosed at its earliest stage.

Scientists were also persecuted: biologists, physicists and scientists of other disciplines. In Leningrad University, the walls of a long corridor were decorated with portraits of professors. When a professor was arrested, his portrait was taken down. The students gave this corridor the name "Arrestometer". Heavy purges struck the Ukrainian Institute of Physics and Technology in Kharkov (the UPTI case): Shubnikov (an outstanding condensed matter physicist), Gorsky and Rosenkevich (co-author of a book by Landau) were executed, Obreimov, Leipunsky and many others were arrested. Out of 400 aircraft designers, about 350 were arrested.

The "Main Enemy" was Trotsky, who had been the People's Commissar for Foreign Affairs at the Brest peace negotiations. Lenin had offered Trotsky the post of Chairman of the Council of People's Commissars. Trotsky declined because he considered that the most senior post in Russia ought not be occupied by a Jew. During the Civil War he was the Chairman of the Revolutionary War Council (up to 1925), and the de facto creator of the Red Army. In his "Testament" Lenin described him as the most talented person in the Party leadership. In 1928 Trotsky was exiled to Alma Ata and in 1929 he was expelled from the USSR. While abroad, Trotsky was the victim of repeated assassination attempts by the NKVD. In 1940 he was killed with an icepick — the murderer was his secretary Mercader, who in reality was an NKVD agent. Later Mercader was awarded the title of Hero of the Soviet Union. The attitude adopted towards "the main enemy" is well represented in Orwell's book "1984". There, a meeting "5 minutes of hate" is held in every office, in which "the main enemy" is shown on a screen and all those present demonstrate their attitude to him by shouting and waving their fists.

The war that Stalin was planning needed some modern arms which the USSR did not have. A huge factory for the production of tanks, called Uralwagonzavod, was built in Nizhny Tagil — the biggest such factory in the world. Similar factories were built in Stalingrad (STZ), Cheliabinsk (ChTZ) and Kharkov (KhTZ) — called respectively the Stalingrad, Cheliabinsk and Kharkov Tractor factories. They did indeed produce tractors, but their technical set-up allowed an immediate switch to the production of tanks. Several aircraft factories were built, the biggest of them in Komsomolsk-on-Amur. Several aircraft design bureaus were created — headed by Tupolev, Miasishchev, Petliakov and others. German, American and other Western technologies were adopted. The closest collaboration was with Germany. An establishment was created near Lipetsk: it comprised a school for training German airmen and a facility for testing new German military planes from Heinkel, Dornier, Junkers (Germany was not allowed to carry out this testing on her own territory under the Versailles agreement). The Lipetsk school continued working up to 1933. A similar school for testing tanks was established near Kazan.

Meanwhile, Hitler continued his offensive. At the Saar plebiscite the majority of the region's population voted for the Saar becoming a part of Germany. (We now know how such referendums are organised.) Austria suffered the Anschluss, which made it part of Germany. Western governments took this in silence. In those regions of Czechoslovakia where the population was predominantly German, a new party (organized by Konrad Henlein) emerged and militated for these regions to be included into Germany. Negotiations were started between Chamberlain (head of the British government), Daladier (head of the French government), Mussolini (head of the Italian government) and Hitler. The tragic result of these negotiations was the signing of the Munich agreement, which allowed the transfer of these regions to Germany. Thereby Czechoslovakia lost its defenses, because the Czech fortifications that protected the country against German invasion were concentrated precisely in these areas. The country had become defenceless, and in 1939 Hitler seized Czechoslovakia.

In February 1936, as a result of elections in Spain, a Popular Front government had been formed, with the participation of Socialists, Anarcho-Syndicalists and Communists. But as soon as in June 1936 a Fascist putsch headed by General Franco started. Germany and Italy gave support to it, sending arms, military personnel and airplanes. German planes bombed Spanish towns which were occupied by Republicans. The town of Guernica was completely destroyed by an air raid. England and France adopted a policy of non-interference. The Soviet Union could only send airplanes with Soviet personnel and a limited military delegation headed by Mikhail Kol'tsov. (He was a journalist but was successful in this field, though it was alien to him. He was later executed, falling victim to Stalin's campaign of terror.) In 1939 the putschists were finally victorious and Franco became the ruler of Spain.

For a while, the foreign policy of the USSR was peace-loving and anti-Fascist. The People's Commissar for Foreign Affairs M.M. Litvinov adopted the slogan "For an indivisible peace, for a collective security". Inside the country, society was also in an anti-Fascist mood. The Soviet Union was a refuge for anti-Fascists who had escaped from Hitler's Germany, newspapers published articles in the same vein, anti-Fascist films were made, such as "The Oppenheim Family", "Professor Mamlok" and others. The word "Fascist" was an insult. The German anti-Fascist singer Ernst Busch, arriving in Moscow, sang:

> Doch ein Mensch, ein Mensch ist
> Hat er Stiefel ins Gesicht nicht gern.[1]

---

[1]But a man is a man, he does not like boots in his face. — (German)

Hitler continued his advance. Now he was demanding that Poland hand over to Germany Danzig and the Polish Corridor which gave Poland access to the Baltic Sea and also cut East Prussia off from Germany. Poland categorically refused this demand. England gave guarantees to Poland — formally promising that in case of a German attack on Poland, England would immediately enter a state of war with Germany. France joined England in this. In summer 1939, Anglo-Franco-Soviet negotiations started. The Soviet delegation, headed by Molotov, demanded that in case of war, Soviet troops would enter Polish territory. Poland categorically refused to accept this condition. (I shall note here that in May 1939 Litvinov had been removed from the post of People's Commissar for Foreign Affairs; he was replaced by Molotov, while Stalin became the Chairman on the Council of People's Commissars.) The negotiations were stopped.

As early as May 1939, the USSR had been conducting secret negotiations with Germany. These were completed on 10 August 1939 by the signing of a trade agreement. On 23 August 1939, a Treaty of Nonaggression was signed with Germany. Ribbentrop, German Minister for Foreign Affairs, came to Moscow to sign this. The agreement comprised secret protocols about the sharing out of Western Europe. At the signing ceremony, Stalin proclaimed a toast to Hitler's health. As a symbol of German-Soviet "friendship", Ribbentrop presented a huge bouquet of red roses to the ballerina Ulanova.

On 1st September, Germany attacked Poland; England and France immediately declared war on Germany. World War Two started. Molotov expressed his feelings about this world war in this sentence: "Let them fight while we will look on and find out how good they are at fighting". And so we did "find out" at the cost of several dozens of millions of human lives.

The tone of Soviet media, and also the general attitude of the population towards Fascism and towards democratic countries changed instantaneously. All critical articles, films and shows immediately disappeared. The newspapers published at the top of pages reports from the German commandment, while reports from English and French sources were published at the bottom and in much less detail. I was most surprised and saddened to see my school friends

(I was in my sixth school year in 1939–40) delight at the news of a German submarine having sunk a British ship.

In 1939–40, Soviet troops entered Western Ukraine, Western Belorussia, Bessarabia (which had been stipulated by the secret articles of the Ribbentrop-Molotov treaty) and also Northern Bukovina. These territories became part of the USSR. An "expression of the will of the people" was organised in Estonia, Latvia and Lithuania, and these countries were incorporated into the USSR as Soviet Republics. At the end of 1939 a war with Finland started, under the pretext that the Finns had attacked the USSR. This war was not a success for the USSR. The Finns offered strong resistance and many Soviet soldiers were killed. In spring 1940 peace was declared, with Vyborg and the territory around Lake Ladoga becoming part of the USSR. But the USSR acquired many enemies in addition to the Finns. The majority of people living in the Baltic region did not welcome their countries being absorbed by the USSR, also the attitude of Western countries towards the USSR seriously deteriorated, with many Communists in these countries leaving the Communist Party. At the early stages of the war on the Western front, during the "Phony War" both Anglo-French and German armies stood facing each other without engaging in any genuine activity. But in April 1940 Hitler suddenly attacked Holland, Denmark and Norway and occupied these countries. In May 1940 this was followed by his strike at Belgium and France. The French army was routed, Paris fell. Only one small zone centred around Vichy remained unoccupied, but it was entirely dependent on Hitler's forces. This zone existed under Marshal Pétain, who had fought well in the First World War, but who this time had capitulated and was carrying out all German orders. The English army found itself squeezed against the sea in the Dunkerque area, but resisted stubbornly and could be successfully evacuated to England. This evacuation was highly important: the South of England had thereby a sufficiently strong military presence, and this did not allow the Germans to land in England under the impetus acquired. But in the words of Churchill — "War is not won by evacuations".

Hitler began preparing a landing on the British Isles. His plan was to gain control of the air space over England and to organise a crossing of the Channel by his armies under cover of this control. On sea, of course, the British fleet was much stronger than whatever Hitler could put together. In the autumn of 1940 the battle for Britain started — a battle of the air. And this battle was lost by the Germans. As Churchill said: "Never had so many owed so much to so few". When he said "the few" he meant the British airmen. The British airmen were much aided by the advanced British science: Britain had radar, which the Germans did not have. Thanks to radar the British could identify German airplanes as soon as these were airborne; they could concentrate British fighter units at the right time in the right place. Moreover, British airmen were able to communicate with each other during aerial battles, but the Germans had no such contact — these radio exchanges had become possible thanks to the quartz stabilization of frequency, invented by British scientists. Radar allowed to locate German submarines which would attack ships bringing armaments from the USA, so many of these submarines were destroyed. Hitler had relied a great deal on a "submarine war" against Britain, but his plans failed. Hitler lost the battle of Britain and relinquished the thought of a landing on the British Isles.

Hitler understood that time was working against him. In the US the production of armament was rapidly increasing, and the weapons produced were reaching Britain. Therefore, in order to provide living space for the Germans, he had to move eastwards. This was a consideration to be added to Hitler's hatred of bolshevism.

Stalin was not afraid of a possible German aggression: the Soviet army was far more powerful than the German one. Stalin knew that in a future war the decisive role would be played by tanks and airplanes. In the western military regions, bordering on Germany, the USSR had immense military power — 14.2 thousand tanks against 4.3 thousand of the Germans and their allies (Hungary, Bulgaria and Finland) and 9.2 thousand airplanes against 5 thousand. Thus, the Soviet army had a 2–3 fold superiority. In the same western military regions, the Soviet army had 3.7 million personnel against 5.5 million on the German side. The German army was somewhat superior in the number of guns: 47.2 thousand against 38 thousand. But the

USSR had huge reserves which could be called up in case of a mobilization. Therefore I share the point of view of Viktor Suvorov, expressed in his book "Icebreaker" [3], that Stalin was expecting to strike first and to wage the war on the enemy's territory. Apart from the western military regions, the USSR had another 10 thousand tanks, 11 thousand airplanes and 40 thousand guns. Germany did not have such reserves: German forces had to be kept in the occupied European countries and on the Western front for fear of a British invasion. A consideration in addition to those presented in "Icebreaker" is that the pacifist and anti-Fascist Litvinov was removed from his posts of People's Commissar for Foreign Affairs and Soviet representative at the League of Nations in May 1939. In other words, it was at this moment that Stalin had taken the decision about a treaty with Hitler, with all its consequences.

I shall write little here about the Fatherland War — much has been written about this. The repressions against the commanders of the Red Army made itself felt from the very beginning of the war. The Fatherland War began on the 22nd of June 1941. The Soviet army was divided into three fronts: North-Western, Western and South-Western, under the command of Marshals Voroshilov, Timoshenko and Budenny respectively. By 10th of July 1941, the losses at the Western front alone were about 12 thousand tanks (only 1700 tanks were left on the Western front) and the losses of personnel were about 2 million people. On 16 July the Germans took Smolensk. After the rout of the armies in the Western sector, Hitler (against the advice of the senior German generals) switched the attack onto Kiev. The commander of the armies in the Kiev sector, General Kirponos, asked Stalin's permission to retreat in order to save the army. Stalin refused. As a result, Soviet armies were surrounded, 600 thousand soldiers were taken prisoner and Kirponos shot himself. On 1st October Hitler's armies started operation "Typhoon" — an advance into the Viazma area. The forces of the Red Army obeyed the orders of Stavka (i.e. Stalin's) — "do not retreat, fight to the last man" — and were surrounded (64 divisions out of 95, 11 tank brigades out of 13, 50 artillery regiments out of 62); irretrievable losses amounted to approximately 700 thousand men [2]. But the weather brought impassable road conditions, so that transporting supplies of fuel and ammunition to the German tank divisions became impossible. Then early winter set in. It was already snowing in Viazma on 8th October. Only 20% of German soldiers had been provided with winter uniforms. Moscow was saved by the winter, with additional help from well-equipped armed forces brought from Siberia. But these tragedies were repeated in spring: in the Crimea three armies were routed while trying a landing over the Kerch straits. Another six armies were surrounded (and taken prisoner) in the Kharkov region. Both operations were under the command of Mehlis, a favourite of Stalin's. The turning point of the war happened only after the victory at the battle of Stalingrad.

Now I shall turn to the most important event of the 20th century, the event that defines our future — the creation of the atom bomb.

2.2. The Road to the Atom Bomb [4–6]; British and American Atom Projects

We may start with Ernest Rutherford's 1911 discovery of the structure of the atom. By bombarding thin layers of various substances with α particles (nuclei of helium) Rutherford discovered that their deflection from their initial direction constituted two groups: small deflections — which were frequent — and large deflections — which happened relatively infrequently. He interpreted this result thus: an atom of size 10^{-8} cm consists of a central nucleus, of radius 10^{-13} cm consisting of protons, and a cloud of electrons which move around the nucleus up to distances of about 10^{-8} cm. Before Rutherford's discovery, the most popular model was the one proposed by J.J. Thompson, in which the atom is seen as a sphere uniformly occupied by electrons and protons. The structure of the atom was understood in 1924 by quantum mechanics, that had been developed by de Broglie, Heisenberg and Schrödinger: in this theory the atomic electrons occupy discrete energy levels. Another source that led to the creation of the atom bomb was the discovery of radioactivity, made by Henry Becquerel in 1896 and investigated by Marie and Pierre Curie and by Rutherford. In radioactive decays, atomic nuclei eject helium nuclei, electrons, positrons, neutrinos, and photons. Rutherford called these particles — helium nuclei, electrons and photons — α, β and γ radiation, and these names have been generally adopted.

In 1931, Joliot and Irene Curie made another step in the study of radioactivity — they showed that radioactivity could be produced artificially by irradiating samples of matter with γ rays. (In 1927 they got married, in 1929 Joliot adopted the name Joliot-Curie and the couple called themselves Joliot-Curie.) But they did not understand that in the case of beryllium the artificial radioactivity is caused not by the γ rays but by neutrons. In 1932 Cockcroft and Walton, irradiating lithium with protons of 600 MeV kinetic energy, managed to split its nucleus into two α particles. This was the first nuclear reaction produced artificially by irradiation of nuclei with protons. (The idea of this experiment was suggested by Gamov.) Already in 1920, Rutherford had proposed the hypothesis of the existence of the neutron, a particle with a mass close to that of the proton and no electric charge. At that time his hypothesis was not accepted. The neutron was discovered by Chadwick in 1932. Experimenting with radiation emanating from beryllium, he could show that its energy was much greater than the energy expected if it had consisted of γ rays. This resulted in a fundamental discovery — the neutron was discovered and it followed that atomic nuclei consist of protons and neutrons, and it removed the contradiction between the atomic number, i.e. the number of protons in the nucleus, and the atomic mass of the nucleus.

The first to propose that, using neutrons, one could get a chain reaction, was Leó Szilárd. Szilárd, a Hungarian Jew, had studied and was working in Germany, but in 1933 because of the racist laws existent in Germany, he moved to Britain. Szilárd gave much thought to a nuclear chain reaction, similar to the chain reaction known in chemistry, which had been discovered by Semenov. It is of interest to remember the circumstances in which this thought struck Szilárd. He was standing in London, waiting for traffic lights to change to green. And suddenly it occurred to him that if a neutron hits a nucleus and causes a reaction in which, on average, there is more than one neutron produced, then a nuclear chain reaction takes place.

The next step in the understanding of the structure of the atomic nucleus was taken by Fermi and his team (Amaldi, Rasetti, Segre and Pontecorvo) in 1934. They irradiated samples of various materials with neutrons. At first, their neutron source was a mixture of polonium with beryllium, and later they started using radon plus beryllium, and they noted something strange happening: the result of the measurement depended on the position of the apparatus — whether it was on the wooden table or in a corner of the room. By chance they had placed a block of paraffin between the neutron source and the sample, and the counting rate strongly increased. That was surprising. After a long discussion of this phenomenon, Fermi suggested a break for lunch. Back in the laboratory, Fermi produced an explanation: paraffin, which contains much hydrogen, strongly slows down the neutrons as a result of their collisions with hydrogen atoms, and slow neutrons have a much greater reaction cross section. (Cross section is a term widely used in physics: it is proportional to the probability of the reaction, and has the dimension of an area.) The physical explanation of this phenomenon was given by Bohr in 1936. Bohr considered the nucleus as consisting of protons and neutrons. When a neutron enters a nucleus, then, as a result of the strong interaction, it scatters frequently off the nuclear protons and neutrons, i.e. it gets "entangled" in the nucleus. Obviously, the smaller the velocity of the neutron, the greater the probability of its "entanglement". In Bohr's theory, this gave rise to the concept of the compound nucleus, and now we know that at low energy the neutron-nucleus reaction cross section behaves as $1/v$, where v is the velocity of the neutron.

Fermi suggested that as a result of neutron capture by heavy nuclei, such as uranium, new trans-uranium elements will be produced. This opinion was contested by the chemist Ida Noddack, who pointed out that after neutron capture the nucleus could break up into nuclei of intermediate atomic weight. As was shown by the further development, they were both right.

The fission of uranium was discovered in 1938 by Otto Hahn and Lise Meitner, joined by the young physicist Fritz Strassmann. Irradiating uranium with neutrons, they noticed that there were many different activities, and they concluded, as Fermi had done, that these activities came from trans-uranium elements. They were working at the Kaiser-Wilhelm Institute in Berlin. Meitner, a Jew but an Austrian citizen, thought that the racist laws did not affect her. But in 1938 Hitler seized Austria and incorporated it into Germany. Meitner had to flee to Sweden. There, with her nephew Otto Frisch, a good physicist, they proved that as a result of neutron capture, the uranium nucleus breaks up, and all the observed activity is the activity of the reaction products, of which there are very many. It was found that in this fission 200 MeV of energy is released. However, the Nobel prize for the discovery of nuclear fission was awarded only to Hahn, but not to Meitner, although she had been nominated 47 times, including 6 times by Max Planck and 3 times by Niels Bohr. Antisemitism had extended itself even into the Nobel Committee, headed at that time by the Swedish physicist Manne Siegbahn.

When news reached Szilárd of the splitting of uranium with large energy release, he realised that there must be neutrons emitted in the process, and if on average more than one neutron is emitted per fission, then a chain reaction can arise and then an atom bomb of vast destructive power can be produced. In a letter to the British Admiralty, Szilárd demanded that his patent of 1936, where the possibility of a neutron induced chain reaction had been described, should be withdrawn, since Szilárd was worried that his discovery would become known to the Germans.

Niels Bohr was the first to realise that fission by slow neutrons occurred with $^{235}_{92}$U, the rarer of the two stable uranium isotopes — only 0.7% of natural uranium is $^{235}_{92}$U, whereas 99.3% is $^{238}_{92}$U. Considering the nearby thorium isotope $^{232}_{90}$Th, which is the only stable thorium isotope, he noted that it has an even number of neutrons — like $^{238}_{92}$U — whereas $^{235}_{92}$U has an odd number of neutrons. Now, since slow-neutron fission of thorium is not observed, he concluded that $^{238}_{92}$U also does not undergo slow-neutron fission.

Later Bohr's statement that ^{235}U was responsible for slow-neutron fission was confirmed experimentally.

Meanwhile, Joliot, von Halban and Kowarsky had published a paper in the April 1940 issue of the journal *Nature*, where they pointed out that 3.5 secondary neutrons are released in fission. Szilárd wrote them a letter persuading them not to publish this paper since Nazi Germany could make use of this discovery to produce an atom bomb, but that letter had no effect.

The decisive moment for the atom project was Einstein's letter to President Roosevelt, which said that an atom bomb could be produced that had an explosive power many orders of magnitude greater than that of conventional weapons. (In fission, about 200 MeV — i.e. $2 \cdot 10^8$ eV — is released per atom, whereas in a chemical reaction, i.e. in explosions of usual grenades or bombs, the released energy per atom is about 1 to 10 eV.) Einstein had been persuaded to write this letter by Szilárd, Wigner and Teller. Alexander Sachs, a lawyer and economist, who had played a significant role in Roosevelt's election campaign, undertook to pass it on by hand to Roosevelt. It is important to note that war had already broken out in Europe: Germany had attacked Poland and was bombing Warsaw; Britain and France had declared war on Germany. Sachs read the letters of Einstein and Szilárd to Roosevelt out loud. From his prior contacts with the President, he knew how to present the message. Roosevelt listened attentively and set into action. He ordered the immediate setting up of a uranium committee headed by Lyman J. Briggs and consisting of physicists and representatives of Army and Navy; he also ordered the allocation of financial resources.

In Britain, the atom project was initiated by Rudolf Peierls, who had emigrated from Germany because of the racist laws. He derived the formula for the critical mass of uranium for the process with fast neutrons. Together with Otto Frisch he found that the critical mass of the isotope ^{235}U is a few kilograms, and the reaction occurs so rapidly that when two pieces of subcritical mass collide, they have no time to separate and the fusion chain reaction takes place.

In Germany, Werner Heisenberg thought that a chain reaction could occur in natural uranium with slow neutrons, if only one could find a suitable moderator. As a moderator one could consider heavy water and graphite. But the graphite would have to be purified of all admixtures, especially of boron, which strongly absorbs neutrons. The German experimental physicist Walther Bothe carried out experiments on neutron capture and came to the conclusion that the neutron absorption in graphite is large and therefore graphite is not suitable as a moderator. (Later it turned out that Bothe's results were wrong, and that graphite can be used. Apparently, Bothe's graphite was not sufficiently pure.) There remained only heavy water — D_2O. In those days, heavy water was produced by electrolysis, and there was only one plant for its production, located in Norway, where there was already a sizable supply available. A German firm attempted to buy the entire stock, but the payment the Germans offered was too low, and the Norwegians refused to sell it. Joliot too wanted to buy the heavy water, and approached the Minister of Armaments, who persuaded the Bank of France to make a credit available, so that the heavy water was bought by France and was made available to Joliot. When Germany occupied Norway, the plant fell into German hands, but Britain sent two Commando units who destroyed the heavy water stored at the plant.

In Spring 1940, the Princeton physicist Louis Turner, comparing the fission of ^{235}U with the absence of fission of ^{232}Th, came to the conclusion that the fission must take place of the transuranium element $^{239}_{94}X$ that is produced by neutron capture of the main uranium isotope ^{238}U, followed by two β decays. In March 1941, Joseph Kennedy, Emilio Segre and Glenn Seaborg produced the proof of this hypothesis in cyclotron experiments, and called this new transuranium element of atomic number 94 plutonium. (The element of atomic number 93 had been previously called neptunium.) That plutonium undergoes fission by thermal neutrons follows from the Bohr-Wheeler theory. From that time, the American atom project proceeded along two parallel paths: the separation of ^{235}U and production of plutonium by neutron irradiation of the main isotope ^{238}U.

From 1941 Britain and the US combined their efforts, and the main work on the atom project moved to America, since Britain was under bombardment by the Germans. The principal participants of the British project — Rudolf Peierls, Otto Frisch, Francis Simon (who suggested passing gaseous uranium fluoride through microporous matter to separate ^{238}U from ^{235}U), and Klaus Fuchs — moved to the US. New people, who up to that time had not taken part in the atom project, joined: Robert Oppenheimer, Hans Bethe, George B. Kistiakowsky and others. It was decided to carry out a criticality experiment using highly purified graphite as a moderator. An underground hall of a stadium in Chicago was chosen as the place for the experiment. Szilárd, using funds made available by President Roosevelt, arranged for the delivery of 400 tons of pure graphite, 40 tons of uranium oxide and 6 tons of metallic uranium. The design was a heterogeneous assembly of blocks of uranium between layers of graphite, so that the fast neutrons would be moderated by the graphite, and the slow neutrons would fall on the uranium. Cadmium rods had been chosen to control the reactor. Fermi led the team running the experiment. The reactor achieved criticality — this was proof of the chain reaction in the natural uranium-graphite system.

In the same year, 1942, Fermi, in a conversation with Teller, expressed a new idea — to use an atomic bomb to start a thermonuclear reaction of deuterium with tritium — or deuterium-deuterium — that occurs at higher temperatures; such a bomb could be made extremely large. Teller was enthusiastic about this idea and started working on it (see below).

On December 7, 1941, Japanese bombers attacked the American Naval base in Pearl Harbor on the Hawaiian islands. Japan had entered the war. At the same time the USA declared war on Germany.

Once a program of action was formulated and it was clear that it could be achieved, it was necessary to create the organizational structure. The atom project was named "Manhattan Project". General Groves, of the Corps of Engineers, was appointed its Head by the Secretary of War. A new laboratory had to be built, sufficient in size and sufficiently isolated for the secret work to be conducted. Robert Oppenheimer was appointed its director — a brilliant physicist and a man of a sharp, fast mind. General Groves considered him a genius (although he did not have such an opinion of, for example, the Nobel laureate Ernest Lawrence, who had built the first cyclotron). A site had to be found for the new laboratory. Los Alamos was chosen, 35 miles north-west of Santa Fe, in a rather desolate area. Laboratory buildings had to be constructed, as well as housing for the staff, electricity and gas supply, communication and roads. Since it was envisaged that far more personnel than at previous experiments would be working at the new Lab (about 10,000), their training for their new job had to be organised. All this was done in less than one to two years. The physical problem that had to be solved was to ensure a complete reaction. For this, one had to prevent the two pieces of uranium or plutonium from detonating prematurely and at the same time from separating before the common explosion. This problem was solved by creating a spherically symmetric implosion, namely by detonating a conventional explosive surrounding the nuclear charge. Such a detonation created the necessary pressure at the inside and satisfied all other requirements. At the center of the nuclear charge was a radioactive source — Po+Be. A huge plant for the production of plutonium was built at Hanford, Washington, and a diffusion plant for the separation of uranium isotopes at Oak Ridge, Tennessee.

Niels Bohr was still in German occupied Copenhagen. In September 1941 he was visited by Heisenberg and von Weizsäcker, who tried to engage him in the German atom project, but he declined. Bohr was half Jewish, and by the racist laws he had to be exterminated. In 1943 he successfully escaped, together with his son Aage, at first by boat to Sweden, and then in the bomb bay of a British bomber to Britain. In the bomb bay he lost consciousness, but recovered on landing in England. The German repressive anti-Jewish laws were introduced in Denmark, but the Danes, with the help of Swedish border guards, succeeded in saving almost all the 7200 Jews who had been living in Denmark. The King of Denmark appeared in the streets of Copenhagen with the yellow star of David sewn on his sleeve. This was prescribed by the Germans racist laws for the Jews, and the King, who was not a Jew, did this to demonstrate his attitude towards these laws.

Bohr went to America and visited Los Alamos. He held the opinion that the atom bomb should not be dropped on a populated area, but on a deserted territory in the presence of representatives of the enemy, to convince them that defeat was inevitable. Afterwards the atomic weapon should be taken under international control, in which the USSR should also take its place. Bohr presented his point of view to important American politicians and also in conversation with Churchill. But his point of view was not shared.

In November 1943 an air-cooled reactor for the production of plutonium was commissioned in Oak Ridge. Then a new problem arose: the reactor produced not only plutonium-239 but also, albeit in smaller quantities, Pu-240. To produce Pu-239, the uranium-238 has to absorb one neutron, and Pu-240 is produced by Pu-239 absorbing another neutron. However, Pu-240 has a far greater spontaneous fission rate than Pu-239, and this could lead to a predetonation. The problem was solved by shortening the duration of the in-core fuel cycle, thereby reducing the uptake of the second neutron. A big plutonium-producing reactor was commissioned on September 26, 1944. However, on the next day this reactor lost criticality and stopped. After some long thought, Fermi and Wheeler understood what had happened: among the fission products, there was iodine with a lifetime of 6.7 hours, decaying into xenon-135 (lifetime 9.1 hours). ^{135}Xe has a huge neutron absorption cross section — it was the neutron capture by ^{135}Xe that had stopped the reactor. After the xenon had decayed, the reactor went critical again. But one had to extract a small amount of plutonium from a great mass of uranium. The US had chemists of high caliber (Kistiakowsky and others) who succeeded in the chemical separation of plutonium from uranium. At the same time, practically pure (95%) isotope ^{235}U was produced in Los Alamos. The time came to test the bomb. The site chosen for the test was near Alamagordo in the southern region of New Mexico. The test was codenamed Trinity. Bunkers with concrete roofs covered with earth were built. To the north, 6 miles from the point of the explosion, instruments were placed, including high-speed cine cameras, to observe the blast. Five miles south of the point were the bunkers of the observers, and 20 miles to the north-east — the place of the VIP guests. One tower was built at Ground Zero — the point of the explosion — another tower 700 meters from it, about 6 meters high. The explosion was to be photographed and recorded on seismometers; the pressure, the optical and nuclear effects were to be registered. Two tanks, lined with lead and closed hermetically, were kept in preparation. The bomb was a plutonium one. The explosion took place at 05:30 on July 16, 1945. The energy of the explosion was estimated by Fermi to be more than 10 kilotons of TNT; later accurate measurements put it at 18.6 kilotons.

On August 6, 1945, a uranium-235 bomb was dropped on Hiroshima. There were 200,000 casualties. On August 9, a plutonium bomb was dropped on Nagasaki — there were 70,000 casualties. The war was over. On August 14, the Emperor of Japan announced in a radio broadcast that he accepted the Potsdam Declaration, i.e. he accepted the unconditional surrender of Japan.

2.3. The American Hydrogen Bomb

Now I turn to the thermonuclear bombs, whose explosive energy is released mainly as a result of the fusion of light elements rather than by fission of heavy ones — uranium or plutonium. I have already mentioned Fermi's idea, developed by Edward Teller, who was later joined by the mathematician Stanislaw Ulam. In 1950 (or possibly 1951) Teller and Ulam came to the conclusion that such a bomb could not be produced. One can take the 31st of January, 1950, as the beginning of the thermonuclear program. This is the date on which President Truman, after the test of the Soviet atom bomb and wishing to maintain American superiority of atomic weapons, signed an order to allocate money for the thermonuclear project. The idea of the thermonuclear bomb is to create external pressure on the trigger device (in addition to the conventional explosive). Then, by the laws of thermodynamics, the temperature of the trigger device increases and reactions can start which do not take place in usual atom bombs. In the first American thermonuclear bomb, *Mike*, a usual atom bomb produced γ rays and neutrons which impinged on liquid deuterium, raising its temperature sufficiently for the neutrons, produced in the reaction $D + D \rightarrow^3 He + n$, to acquire an energy above the fission threshold of ^{238}U. They impinge on ^{238}U and initiate the explosion of the entire device. *Mike* was installed on the Enevetok Atoll and detonated on November 1, 1952. The energy of the blast was 10 Megatons. But this device could not be transported: its weight was 82 tons and it could not be delivered by an aircraft. This shortcoming was removed with the second device *Bravo*, detonated on March 1, 1954. It was a two-stage device; the pressure was created by the γ rays from the first stage — an atom bomb — and directed onto the second stage that contained ^{238}U. The energy of *Bravo* was 15 Megatons; substantial buildings were destroyed within a diameter of 8 kilometers.

2.4. The German Atom Project [6, 37, 40, 41]

The main work on the German atom project was carried out at the Kaiser Wilhelm Institute of Physics (KWI), where in 1938 Otto Hahn, Lise Meitner and Fritz Strassmann carried out experiments on neutron absorption in uranium. An unexpected result was that the irradiated specimens contained nuclides of intermediate atomic weight. At first the authors assumed that these nuclides had arisen by the decays of transuranium elements. However, after Lise Meitner's emigration to Sweden, she, together with Otto Frisch, showed that the uranium nuclei themselves were fissioning,[2] and that the nuclides of intermediate atomic weight were the products of this process. Early in 1940, Frédéric Joliot, Hans Halban and Lew Kowarski found that the mean number of neutrons emitted in fission is 3.5, establishing the possibility of nuclear chain reactions and of production of huge amounts of energy. The German Ordinance Research Department (Heereswaffenamt) took an interest in the question of nuclear fission. Its Head was Colonel Erich Schumann; he ordered the KWI for Physics to be mobilised for war-related work. Kurt Diebner was appointed its formal Head, but the actual work on the atom project was headed by Heisenberg, whose group included Carl Friedrich von Weizsäcker, Karl Wirtz and Fritz Bopp. In his classified report "Possibilities of technical energy production by uranium fission" presented in December 1939, Heisenberg stated that the fissile isotope of natural uranium is U-235, and its content is 0.7%. (He knew already of Bohr's work on this subject.) The reactor studied was working on thermal neutrons, because their cross section is much greater than that of fast neutrons. Due to a wrong result found by Walther Bothe,

---

[2]The word *fission* for this process was coined by Otto Frisch.

who had been studying neutron absorption in graphite, graphite was rejected as a moderator, and the only moderator left was heavy water. But the amount of available heavy water was small. In 1940, Paul Harteck (in Hamburg) and Fritz Houtermans[3] (in the von Ardenne group) showed theoretically, using the Bohr-Wheeler model of the nucleus, that the element of atomic number 94 and mass number 239 (plutonium) was fissile by thermal neutrons. Plutonium is produced by absorption of thermal neutrons by uranium-238, followed by a sequence of two beta decays. That implies that plutonium can be produced in reactors working on thermal neutrons.

Heisenberg later maintained that their aim was to create a reactor for propulsion. As a result of Anglo-American bombing, it was becoming more and more difficult in 1943 to work in Berlin. The work was moved to Hechingen in Württemberg. The last experiment was carried out in February 1945 in a cave in the town of Haigerloch. It showed that a self-sustaining chain reaction was very near but was not yet achieved. A short time later American troops entered the German nuclear center. Heisenberg, von Weizsäcker and others were taken into custody by the ALSOS Mission. They were interned in a mansion in England at Farm Hall, where all their conversations were recorded by eavesdropping equipment. The German scientists were greatly surprised when they learned that the Americans had produced an atom bomb — they thought that this would need several decades of work. Houtermans was not interned. He moved to

---

[3] Houtermans' fate was unusual [15]. He was an anti-Fascist. When Hitler came to power, he emigrated to the Soviet Union and worked at the Kharkov Institute of Physics and Technology (KIPT). This institute was founded by A.F. Ioffe, and it comprised a strong group of physicists: Landau, Shubnikov, Rosenkevich, Gorsky, Leipunsky, Sinelnikov and others. In the 1930s, the institute was host to international conferences on nuclear physics. But in 1937, many physicists were arrested (the KIPT case), and some of them were shot (Shubnikov, Gorsky, Rosenkevich). Landau was not arrested: shortly before this action he had moved to Moscow and worked at Kapitsa's Institute for Physical Problems. Houtermans was also arrested and was imprisoned until 1940, when during the time of friendship with Germany he was handed over to the Gestapo. Comparing the investigators of the NKVD and Gestapo he concluded that those of the Gestapo were less professional than those of the NKVD. For instance, Houtermans was sitting across the desk of his investigator who did not realise that he was able to read upside down. From the Berlin prison, he could let von Laue know that he was in Berlin. Von Laue realised that he was in a prison and succeeded in getting him released and join the group of Manfred von Ardenne.

Switzerland, where he set up an Institute for Theoretical Physics at the university of Bern.

2.5. The Soviet Atom Project [5, 9, 16–22]

In the USSR, nuclear physics research began in 1932. As mentioned above, this was a year of a breakthrough in the study of nuclear physics. Abraham Fedorovich Ioffe had founded the Institute of Physics and Technology (Phystech) in Leningrad. He also tried hard to ensure that research in physics be pursued all over the country (he established, for instance, the Ukrainian Institute for Physics and Technology and directed a number of researchers to it). Ioffe could not ignore any achievements in physics, and organised a group for nuclear physics at the Phystech, which he himself headed. But after half a year he appointed Igor Vasilievich Kurchatov as Head of the group, which was then named the Nuclear Physics Division. Kurchatov, who had been working up to that time on ferroelectricity, moved vigorously into the new subject. His first — but famous — papers on nuclear physics were on the physics of isomeric (i.e. metastable) nuclear states.

Another group in the same institute was headed by Abraham Isakovich Alikhanov. This group was part of the division headed by Lukirsky. They studied processes of internal and external electron and positron conversion in nuclear β decays. Conversion is a process of ejection of an atomic electron by a γ quantum emitted by its nucleus. In 1935 Robert S. Shankland, studying Compton scattering of γ quanta on electrons, came to the conclusion that energy was not conserved in this process. Many physicists, among them Niels Bohr, believed this result, and Bohr even advanced the hypothesis that energy is not conserved in atomic processes, and that energy conservation arises only statistically, i.e. requires many particles. In 1936, Alikhanov, Alikhanyan and Artsimovich set up an experiment for the annihilation of slow positrons: $e^+ + e^- \rightarrow \gamma\gamma$. In this experiment, if energy is conserved, then the γ quanta must fly apart back-to-back, and each one must have exactly the energy mc^2, where m is the electron mass and c is the speed of light. This experiment demonstrated the conservation of energy in an elementary process. Alikhanov was appointed Head of a section of Lukirsky's Division and also was elected Corresponding Member of the Academy of Sciences of the USSR.

Uranium was studied in Leningrad not only at the Polytechnic Institute, but also at the Radium Institute, where such well known scientists as Vernadsky, Khlopin and others were working. At the initiative of Vernadsky, the Uranium Commission of the Academy of Sciences was established. He insisted on a search for uranium deposits and on other investigations on uranium to be undertaken.

A.F. Ioffe organised conferences in the USSR on nuclear physics to which foreign scientists were invited. The first of these conferences took place in 1933. Paul Dirac, Frédéric Joliot-Curie, Franco Rasetti, and Victor Weisskopf took part in this conference. Several such conferences were organised later on.

Before the war, the Radium Institute had a cyclotron — a proton accelerator — built by a group headed by Mysovsky. But that accelerator was rather unstable. Therefore Kurchatov and Alikhanov decided in 1939 to build a new cyclotron in the Radium Institute, with a higher energy and stable. The construction of this cyclotron was nearly finished, but work was disrupted by the war. At the Kharkov Institute of Physics and Technology, Sinelnikov and Leipunsky repeated in 1936 the work of Cockcroft and Walton.

In 1939, Zel'dovich and Khariton wrote three papers on the possibility of using uranium for energy production and on the development of the chain reaction in uranium. They came to the following conclusions: (1) a chain reaction is not possible in natural uranium either by fast or by slow neutrons, and therefore natural uranium is not suitable as a source of energy; (2) in the isotope uranium-235 a chain reaction with slow neutrons is possible, and its mass must be of the order of 10 kg (in agreement with the result of Peierls and Frisch); as a moderator in the uranium-moderator system, only heavy water is suitable, and not graphite [9, 16–22].

Georgy Flerov was the first to learn about the western atom project. Some time before the war, he and Konstantin Petrzhak had made a discovery — they observed the spontaneous splitting of uranium nuclei. In 1942 Flerov was serving in the army. Having some free time, he visited the library of the University of Voronezh to see what was new in nuclear physics. He noticed that, whereas previously there had been many papers on this field, now there were none. He concluded that work on nuclear physics was now classified and that a war-time project was being carried out. He wrote about his observation in a letter to Stalin, but did not receive a reply. After some time he wrote again to Stalin, developing the "gun-scheme" of using uranium for an atom bomb. But again he did not get a reply. At about the same time, a notebook was found in the pocket of a dead German officer containing information about the German atom project. Colonel I.G. Starinov sent this information to Stalin, but there was no response. A response came when in 1943 Beria presented to Stalin documents obtained by the intelligence service in England and the US (before that time Beria had not passed on this information). The information reported: (1) an atom bomb made with uranium-235 could be produced, and its weight would be about 10 kg; (2) the most efficient method of separating uranium-235 from uranium-238 was the diffusion method of forcing gaseous uranium hexafluoride through semi-permeable membranes. The explosive energy of such a bomb of 10 kg weight would be equivalent to 1,800 tons of trinitrotoluene (TNT); in addition there would be a fall-out of a large amount of radioactive material. The cost of building a plant to manufacture 3 bombs per month was estimated as 5 million English pounds. The time needed to produce an atom bomb would be about 3 years; thus, it could be produced during the war, and it would have a decisive effect on the conduct of warfare.

After hearing Beria's presentation, Stalin decided to form a group of physicists to investigate the problem of producing an atom bomb. He instructed Molotov and Kaftanov (the Head of the Higher Education Committee that was in charge of questions of science in the government) to find the Head of such a group. Molotov and Kaftanov approached A.F. Ioffe, a Vice-President of the Academy of Sciences of the USSR, and a well known physicist. In reply, Ioffe wrote that this task would take many years, and he was too old to head such a group (he was 63 years old then). He proposed as candidates for the role of Head of such a group his PhD students Alikhanov and Kurchatov (in that order). Molotov and Kaftanov invited both (they were outside Moscow) and interviewed them. Kurchatov made a good impression on them, Alikhanov — a rather poor one. In 1943, Laboratory No. 2 was set up within the Academy of Sciences, and I.V. Kurchatov was appointed its Head. At first the laboratory was quite small: it consisted of Alikhanov, Zel'dovich, Khariton, Kikoin, Pomeranchuk, Flerov and a few more people. Pomeranchuk later remembered: "The entire Laboratory staff fitted into Kurchatov's jeep". The main thrust of the assignment was the separation of isotopes U-238 and U-235.

After receiving intelligence reports (mainly from England) that element eka-osmium $^{239}_{94}$X, called plutonium, was also fissile by neutron capture, Kurchatov abruptly changed the main direction of research: it was now directed towards the production of plutonium in thermal neutron nuclear reactors containing natural uranium and a moderator that could be either graphite or heavy water. Information on the possible use of heavy water as a moderator had been received from Klaus Fuchs. (Briefly, this possibility was contained in a paper by Zel'dovich and Khariton.) Fuchs, a German and a Communist, had emigrated to England after Hitler's accession to power. He was invited by Peierls to join his group. Fuchs showed himself as a very able physicist, but he thought that if the US and Britain were to have an atom bomb, then the Soviet Union should also have such a bomb. After Peierls' group, including Fuchs, had moved to the US and joined the Manhattan Project, Fuchs made contact with Soviet agents and passed on to them all his information. The Soviet Union received details of the first nuclear reactor on natural uranium with graphite moderator that was started by Fermi in 1942, a blueprint of an industrial reactor for the production of plutonium — natural uranium with graphite, a blueprint of a Canadian reactor — natural uranium and heavy water. (Until 1943 Beria did not pass on this information to Stalin, neither did he consult physicists — incompetence had left its mark even here.)

Alikhanov favoured the use of heavy water instead of graphite for slowing down neutrons, because heavy water absorbs fewer neutrons than graphite does. Moreover, the use of heavy water as moderator allows the prevention of a positive temperature effect, which was a great danger in a graphite reactor. Laboratory No. 3 (now the Institute of Theoretical and Experimental Physics, ITEP) concentrated on designing and building heavy water reactors, in other words reactors in which the neutrons are slowed down (*thermalized*) by heavy water. This became the Laboratory's main activity for many years. Lab. No. 3 was the second Soviet laboratory of the atom project. The blueprint of the Canadian reactor with heavy water moderator and also samples of uranium-235 and uranium-233 (uranium-233 is not stable but it lives for a very long time and can be used for an atom bomb) were handed over by the British physicist A.N. May.

Material obtained by intelligence services would be handed on to Kurchatov exclusively. He could write on the document "For information to ..." and give one or at the very most two names. These people, though, could not refer to the source of the information received and were obliged to present these results as their own. A drawing of the atom bomb was also handed over. In 1946 Fuchs first informed the USSR briefly that a thermonuclear bomb was possible, and later sent a detailed description of it (see below). Work on the atom project was however rather slow and engaged few people until 1945, when news of the atom bomb exploding over Hiroshima arrived. (First, Truman told Stalin at the Potsdam Conference about a weapon of extraordinary strength, and information about the explosion itself came later.) This speeded up the USSR's joining the war against Japan, since Stalin was endeavouring to seize as much territory as possible before Japan capitulated.

After the explosion of the atom bomb over Hiroshima, work on creating an atom bomb in the USSR developed very fast. A Special Committee was organised on 20th August 1945, headed by Beria. It comprised prominent figures of Party and government and also two scientists: Kapitsa and Kurchatov. A Scientific and Technical Council began to work. Vannikov was appointed chairman and Alikhanov its Scientific Secretary. Stalin summond Kurchatov and said: "If the child does not cry, the mother does not know its need. Ask and you will be given all you need". The 30th of August 1945 saw the creation of The First Main Directorate (Russian initials PGU), of the USSR Council of People's Commissars which was entrusted with the day-to-day administration of the atom industry. B.L. Vannikov was appointed its Head, and he was also the Vice-Chairman of the Special Committee. A.P. Zaveniagin, Vice-People's Commissar for Internal Affairs, was appointed deputy to Vannikov. In September the Technical Council of PGU held a meeting which heard a series of presentations:

1. I.V. Kurchatov and G.N. Flerov (each presented his own paper): "The production of plutonium in uranium-graphite reactors with ordinary water cooling";
2. A.I. Alikhanov: "The production of plutonium in natural uranium reactors with heavy water as moderator and coolant";
3. I.K. Kikoin and P.L. Kapitsa: "The production of enriched uranium by the gas-diffusion method";
4. L.A. Artsimovich and A.F. Ioffe: "Enrichment of uranium by the electromagnetic method".

Alikhanov was appointed Head of Lab No. 3; it was given the main building of the pre-revolutionary estate of Menshikov in Cheremushki (built in 1780). The following tasks were assigned to the Laboratory:

1. Physics research, design and construction of a uranium-heavy water chain-reacting pile;
2. Physics research on the systems thorium-water and thorium-plutonium-water for the production of ^{233}U, which fissions by absorption of slow neutrons;
3. Physics research on β radioactivity;
4. Physics research on high energy nuclear particles and cosmic rays.

Soviet scientists made great progress in the theory of thermal neutron nuclear reactors. They primarily studied heterogeneous systems, in which cylindrical slabs or plates of uranium were placed in the moderator, because they understood that in this case the probability of loss of neutrons due to absorption by U-238 resonances would be less. Still in 1940, Zel'dovich and Khariton defined the concept of the neutron reproduction factor k — the ratio of the number of neutrons produced in fission to the number of absorbed neutrons for an infinite medium where one can neglect the loss of neutrons due to escape from the system. For k they found the formula

$$k = \nu' \varphi \theta \varepsilon$$
$$\nu' = \frac{\sigma_f}{\sigma_c + \sigma_f}$$

where ν is the number of neutrons produced in fission by absorption of one thermal neutron, σ_f is the fission cross section, σ_c is the cross section of absorption of a neutron by fissile material without fission (all cross sections are for thermal neutrons), φ is the probability of a fast neutron produced in fission to reach thermal energy without being captured by a resonance, θ is the probability of absorption of a thermal neutron by fissile material. The coefficient ε, that was introduced later by Flerov, takes account of fission by a fast neutron in the same uranium slab. The necessary condition for a sustained chain reaction is

$$k > 1$$

The most complicated problem is the calculation of φ. This problem was solved by Gurevich and Pomeranchuk in 1943 [17]. Basically their theory is the following. Consider some level (for instance, the ^{238}U level at $E_r = 6.67$ eV of width $\Gamma = 25 \cdot 10^{-3}$ eV and a peak cross section of $\sigma_0 = 22 \cdot 10^3$ barn, 1 barn $= 10^{-24}$ cm^2). Define the width of the "danger zone" ΔE_r by the equation

$$\sigma_a \left(E_r + \frac{1}{2} \Delta E \right) \rho \bar{l} = 1$$

where $\sigma_a(E)$ is the absorption cross section, E_r is the neutron energy at the peak of the cross section, ρ is the number of absorbed atoms per cm^3 of the slab, and $\bar{l}$ is the mean free path of the neutron in the slab. Neutrons of an energy at the center of the resonance level are absorbed with 100% probability by the uranium slab. The equation shown corresponds to such a definition of the "danger zone" for which the probability of absorption of the neutron by the uranium slab is

50%. Assume that near the resonance $\sigma_a(E)$ is described by the Breit-Wigner formula

$$\sigma_a(E) = \sigma_a(E_r)\sqrt{\frac{E_r}{E}}\frac{1}{1+x^2}, \qquad x = \frac{2}{\Gamma}(E - E_r)$$

Substitute this equation into the previous one, hence $\Delta E = \Gamma\sqrt{a}$, where $a = \sigma_a(E_r)\,\rho\,\bar{l}$. For the first — the most dangerous — level of ^{238}U $a \gg 1$, hence $\Delta E_r \gg \Gamma$. This means that the dominant contribution to the resonance absorption is from the tails of the resonance. If we separate the resonances in two groups: low lying and high lying, then the discussion presented is appropriate for the former one. For the latter group the resonance absorption is just proportional to the number of atoms. The final formula for the resonance absorption derived by Gurevich and Pomeranchuk for cylindrical slabs is [17]

$$-\ln\varphi = \frac{\lambda_s}{\xi}\frac{\alpha\,d^{3/2} + \beta d^2}{a^2 - \frac{\pi d^2}{4}}$$

where α is expressed in terms of the resonance parameters, λ_s is the scattering length in the moderator, ξ is the mean logarithmic energy loss in the moderator, $1/\xi \approx A/2 + 1/3$. This result is valid for small slabs with $\bar{l} < \lambda_s$. In the US the resonance absorption was considered by E. Wigner, who proposed the following interpolation formula:

$$-\ln\varphi = \frac{\lambda_s}{\xi}\frac{A\,d + Bd^2}{a^2 - \frac{\pi d^2}{4}}$$

The formulæ differ in principle ($d^{3/2}$ in the former and d in the latter). In the early stages of the atom projects in the US and USSR, when the parameters of the levels of uranium were not yet known, the constants in both formulæ were determined empirically, and the fits gave in both cases satisfactory results. However, even in this case the formula of Gurevich and Pomeranchuk has an advantage over Wigner's formula: (1) it predicts the Doppler broadening of the level, i.e. the magnitude of the temperature effect. With increasing temperature the curve of the resonance absorption changes in the following way: the peak cross section gets smaller, but the area under the curve does not change, i.e. the tails are lifted. The resonance absorption of the neutrons increases, since it is determined by the tails of the resonance, i.e. φ decreases. (2) it allows one to do the calculation in those cases when the moderator is inside the slab, which was done by Rudik at ITEP.

In 1947 Akhiezer and Pomeranchuk wrote a monograph "Introduction to the Theory of Neutron Multiplication Systems (Reactors)", where a detailed theory of nuclear reactors was presented [18]. In addition to the theory of the resonance absorption, they considered the determination of the critical size, worked out the basis of the theory of heterogeneous reactors, studied the theory of the reactor kinetics and the temperature coefficient, and gave an exact solution of the boundary value problem for a flat boundary between the medium and vacuum (by way of solving Boltzmann's equation using the Wiener-Hopf method). That was the first book on reactor theory. (In the US a book on reactor theory appeared 3 years later and was not as good.) This book was classified and was published only in 2002. Other ITEP scientists also made significant contributions to reactor theory: Galanin developed the theory of heterogeneous reactors [19, 20], Galanin and Ioffe computed the cycle Th $-^{233}$ U for the heavy-water moderator [21] and showed that it can be used as a breeder; Ioffe and Okun developed the theory of deep burn-up in heavy-water reactors [22].

In 1945 Kapitsa repeatedly wrote to Stalin, protesting against the secrecy imposed on the atom project and against the fact that the Special Committee was run by incompetent people, such as Beria, Malenkov and Voznesensky. He proposed that we follow our own path towards the creation of the atom bomb, different from that taken by America. Kapitsa maintained that there cannot be a Soviet science or an English science but only one science — the international science. In this respect, his views were similar to those of Bohr. Bohr had written to Kapitsa that scientists ought to meet in order to discuss the consequences of the atom bomb having been created. Kapitsa supported this proposal. Instead of arranging such a meeting, Beria sent Yakov Terletsky to Copenhagen to meet Bohr. Terletsky was a science adviser to "Department C" of NKVD, he lectured on statistical physics at Moscow university and hardly knew anything about nuclear physics. The aim of Terletsky's visit was to worm information about the bomb out of Bohr. But most of Bohr's answers to Terletsky's questions carried little information. One answer, however, was indeed worthy of interest and might have provided some information which would have been useful at the time. Terletsky asked Bohr after what time the uranium slabs containing plutonium were taken out of the nuclear reactor. Bohr said that he did not know exactly, but he rather thought it was after one week. Bohr's answer was entirely wrong — maybe he genuinely did not know, maybe he intentionally gave Terletsky an incorrect duration. (In fact, the answer should have been "after about a month".)

Terletsky's visit to Bohr was one of a series of episodes of the opposition between Kapitsa and Beria. On 19th of December 1945 Kapitsa handed in his resignation from the atom project. Beria asked Stalin to allow the arrest of Kapitsa, but Stalin refused, saying: "I shall take him out of your way myself, but you leave him alone". In August 1946 Stalin signed an order removing Kapitsa from the post of the Institute's Director. Kapitsa was forced to carry out research at home, at his dacha on Nikolina Gora, but there was an NKVD post very near by.

Work on creating an experimental graphite-uranium reactor began in 1945. To a considerable extent, Kurchatov used the drawing of the American reactor. (For instance, the uranium rods had the same diameter as the American ones.) The work was hampered by a shortage of uranium and graphite of the necessary purity. (I would like to remind the reader that it was precisely because the graphite was not sufficiently pure that Bothe in Germany had come to the conclusion that graphite was not suitable as a moderator.) Nonetheless, on the 20th of December 1946 the experimental reactor F-1 was started, the chain reaction was activated. The power of the reactor was no higher than 200 W, it was suitable only for experimental research, and not even for all of it. The reactor worked without cooling.

The first industrial reactor was built in the Urals, 15 kilometers east of the town Kyshtym, near lakes which could provide cooling water for the reactor (Cheliabinsk-40). The buildings for the reactor with all attendant services and a new town were built by inmates of labour camps. They also mined the uranium ore and built part of Labs 2 and 3. No one looked after their health, although they were exposed to radiation. Therefore mortality was very high among these prisoners. On 22 June 1948 the reactor was started, its power was 100 MW. The same difficulties had to be overcome here as those faced previously by the Americans: poisoning by xenon, the need to restrict the fuel-cycle due to ^{240}Pu accumulation etc. But the intelligence service had provided information about ways to overcome all these. A factory was built for the chemical extraction of plutonium from uranium. By 1949 the first kilograms of plutonium had been produced.

At the same time, Lab 3 was working on the heavy-water alternative. This necessitated first of all the production of heavy water — there was practically none of it in the USSR — and then an experimental heavy water reactor could be built. Heavy water was produced by electrolysis at the Chirchiq industrial complex. A heavy water reactor is very different from a graphite one, it is a complex physical installation. Heavy water reactors are different in that the neutrons produce an explosive mixture of hydrogen and oxygen which needs to be removed and burnt. This type of reactor was built using a Canadian design, but with one important difference: the lid of the reactor could be rotated. This allowed the possibility of changing the spacing of the lattice, thereby one could choose both the very best size of uranium slabs and the lattice spacing, and at the same time one could check the accuracy of theoretical calculations. One of the advantages of the heavy-water reactor was that it needed less than 10 tonnes of uranium, whereas a graphite reactor needed approximately 150 tonnes. The reactor was commissioned in April 1949, its power was 400 kW.

In 1955, ring slabs were introduced in this reactor, 2% enrichment, its power increased to 2.5 MW and the neutron flux in the center of the reactor was $4 \cdot 10^{13}\,\text{s}^{-1}\text{cm}^{-2}$. Improved 10 MW reactors of this type were built to the ITEP design in China and Yugoslavia. In 1964, Yu.G. Abov and P.A. Krupchitsky using this reactor discovered parity violation in strong interactions. Later several industrial heavy-water reactors were built and successfully exploited in Cheliabinsk-40.

The American project of detonating a plutonium bomb was chosen. Initially, the designer of the first bomb was V.A. Turbiner. He introduced several improvements to the American design and insisted on them. But Stalin said: "Everything must be done exactly as in the American design!" Turbiner was dismissed and replaced by N.L. Dukhov, who had previously been director of some military factory. Dukhov knew nothing about nuclear physics and understood very little about the construction of an atom bomb, as he had no experience in this matter. Stalin, however, insisted on appointing him to the post of principal designer of the bomb. After its successful test Dukhov was rewarded with the title of Hero of Socialist Labour, while Turbiner was awarded a monetary bonus: an addition of one month's salary. The test was carried out on 29 April 1949 at a testing range 160 km from Semipalatinsk. Two three-storey buildings were built on the test site to record the effects of the explosion, a tower for the bomb, a substantial building for the Control Center and some others. The power of the explosion was approximately 20 kT of TNT. As he was congratulating Kurchatov on the success of this test, Beria embraced and kissed him, saying "There would have been much dreadful trouble if this had not worked!" Kurchatov knew very well what sort of dreadful trouble there would have been.

2.6. The USSR's Thermonuclear Bombs [1, 5–9, 11]

The very first system of a thermonuclear explosion that could be practically realized belonged to Sakharov. The explosive device which Sakharov nicknamed "layer cake" ("sloyka" in Russian) consisted of alternating layers of uranium-238 and liquid deuterium with a quantity of tritium added (as heavy water). The fuse consisted of an ordinary atom bomb which created pressure and thereby increased the temperature and the neutron energy sufficiently for fission of uranium-238. Later on, V.L. Ginzburg proposed an improvement of the system, replacing deuterium by ^{6}Li D. Then the neutrons, impinging on the deuterium, split it without practically any energy loss: $D + n \to p + n + n$ (A). If, on the other hand, the neutron hits the ^{6}Li, then the reaction is $^6Li + n \to T + {}^4He$ (B). But the cross section of the reaction $D + T \to {}^4He + n$ is two orders of magnitude greater than the cross section of $D + D \to {}^3He + n$, and the temperature at which this took place was substantially lower. Thus, in both cases — processes A and B — there was a substantial advantage. The testing of this bomb was carried out at the Semipalatinsk test range on 12 August 1953. The TNT equivalent of the explosion was 400 kt, which means that it had 20 times the power of the bomb dropped on Hiroshima. The bomb could be transported, i.e. it could be used as a thermonuclear weapon. The radius of impact, that is the distance from the epicenter to the boundary within which the temperature rose to 300 centigrades, was of the order of 2.5 kilometers. The next step in the development of the hydrogen bomb was the use of γ rays to produce pressure in the explosive part of the bomb. To achieve this, the bomb was enclosed in a heavy metal casing. Its explosive material was ${}^6Li\, D$; tritium was not used. A half-size version was dropped from an aircraft on November 6, 1955. The TNT equivalent of the full-size bomb was 3 Mt. This development was strongly pursued, and in 1961 another half-size version was dropped from an aircraft on the test site of Novaya Zemlya; its TNT equivalent was 50 Mt (therefore the full-size bomb would have 100 Mt TNT). As a result, the entire North of European Russia was contaminated by radioactive fall-out, and even in Moscow the radiation rose to several times above background.

2.7. Why the Soviet Union Needs Nuclear Weapons

The usual answer to this question is: to resist an American nuclear attack or to resist American nuclear blackmail (I confess that I never heard any other answers). But this answer is wrong, as will be shown below. The first atom bomb was tested in the USSR on 29 August 1949. It was the one and only in the land, there were no more atom bombs in store. This was a plutonium bomb, the plutonium for it was produced in just one reactor and the production capacity of this reactor was of the order of 10 bombs per annum.

In the US at that time (November 1949), according to the plan of the United Committee of Heads of Staff (the highest military organ in the US) it would be necessary to have 130 atom bombs of the 20 kt Hiroshima type to achieve the full destruction of the USSR's military and economic potential. By the end of 1949 the US already had this number of atom bombs at their disposal and also enough airplanes to deliver them (these were 35 B-50 bombers, equipped for the transport of atom bombs from US territory with landings on American bases in Europe or in the Middle East or Far East, 36 B-36 from US and allied bases and more than a hundred B-29). All these planes could fly at heights over 40 000 feet, beyond the reach of Soviet anti-aircraft guns and interceptor fighter planes. In addition, the USSR did not have any airplanes capable of reaching US territory. Stalin set the task of creating such a plane to Tupolev, but Tupolev refused, saying that he could not build such a plane. Stalin set the same task to Miasishchev who spent two years working on it but did not succeed.

Therefore at the end of 1949 the Soviet Union was completely defenseless against a nuclear attack from the US. But a nuclear attack did not happen.

By the time of the Korean war (summer 1950), when American and South Korean forces were pressed against a narrow strip of the Southern coast of the Korean peninsula, General McArthur (Commander of the armed forces in the Far East) and Admiral LeMay (Commander of US Air Force) discussed the question of using atomic weapons in Korea (at that time, the US had an even greater advantage over the USSR as far as atom bombs were concerned), but they rejected their use without even talking to Truman — the use of atomic weapons could only be allowed by the President of the United States. They preferred the option of sending a landing force to Inchon.

The situation changed at the end of the 1950s–early 1960s. Although the US continued to retain its advantage in the number of atom bombs (18,000 in 1961), and although their number greatly exceeded any need in the case of a possible war with the USSR, the situation changed because the USSR had developed ballistic missiles capable of reaching the US. (I would remark here that the increase of energy produced by the explosion of a bomb from 10 Mt to 100 Mt is inefficient, because the effective thickness of the atmosphere is of the order of 10 kilometers, and the destructive radius of a 10 megaton bomb is of the same order). Stalin, who had put forward the explanation which was quoted earlier, could of course not foresee how atom bombs would develop over the next 10 years, so that this explanation was wrong straight from the beginning.

Nowadays, due to the treaties START-1 and START-2, the USA and Russia have approximately 8,000 ballistic missiles with nuclear warheads. The power produced by one American warhead would be of the order of 2 Megatons, one may think that the power of Russian ones is similar. Figures quoted earlier show clearly that both the number of warheads and their explosive energy exceed the necessary level by several orders of magnitude, even in the case of a war between Russia and the US, and even after anti-missile defense has been taken into account. (It is possible to intercept a missile only at the initial stage of its trajectory, when it starts to gain speed, because from then on its warhead follows a ballistic trajectory and becomes invisible in practical terms.) Therefore I would consider it sensible to limit all nuclear countries to approximately 10 ballistic missiles with nuclear warheads, and even then missiles and warheads should be kept apart from each other. This would reduce the possibility of an unsanctioned launch or of a failure of controlling apparatus etc.

I would prefer that nuclear arms did not exist at all or — since they do exist — that they were completely destroyed — if I was not afraid that a wild dictator with an atom bomb would come to power in some country. Nonetheless, I approve the bombing of Hiroshima and Nagasaki. These bombings led to the death of 270 thousand people (including those who died as a result of radioactive irradiation after the bombing). But the war stopped: the bombs were launched on 6 and 9 August 1945, and on 14 August of the same year the Emperor of Japan broadcast his order of unconditional surrender of the Japanese Armed Forces. By that time the Japanese army had occupied the Malayan archipelago, Burma, Indochina, a considerable part of China, Manchuria, Korea, and some islands in the Pacific Ocean. If the war had continued, they would have needed to be dislodged from all these territories by force. An American landing on

the Japanese islands would have been impossible to avoid, also Soviet troops would inevitably been engaged in fighting for Manchuria,
Southern Sakhalin and the Kurils. All this would have led to even
more loss of life. (For instance, before the capitulation of Japan 90
thousand Japanese and 30 thousand American troops lost their lives
in the battle of Saipan, an island of size 120 square km.) The greatest
human losses would have been incurred in Japan itself, where civi-
lians, including children, would have been killed as well. Therefore,
the decision to launch atom bombs over Hiroshima and Nagasaki,
which did initially appear barbaric, was in fact a humane act: it did
reduce the number of casualties. The blame for these deaths must be
laid on the leaders of Japan who had started the war in the Pacific
Ocean.

2.8. The Fundamental Cause of the Chernobyl Disaster [20]

The fundamental cause of the Chernobyl disaster was a positive temperature coefficient of a reactor. The temperature coefficient is the
rate of change of the reactor power with changing temperature. The
Chernobyl reactor had water coolant and graphite moderator. For
simplicity, consider the cross section of the fuel rod of the reactor as
shown in Fig. 1. The uranium is enclosed in a thin layer of aluminium, and another layer of aluminium separates the coolant water

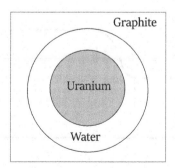

Figure 1: Cross section of a fuel rod in our model. (In the Chernobyl reactor the fuel
rods were more complicated, but the conclusions are the same as for our model.)

from the graphite. (In the calculation the layer of aluminium can be neglected.) The neutron distribution is described by the diffusion equation with boundary conditions

$$D\frac{\partial n}{\partial r}\bigg|_i = D\frac{\partial n}{\partial r}\bigg|_k, \qquad D = \frac{1}{3}l_t V$$

at each boundary (i, k labels uranium, water, and graphite, l_t are the transport lengths, $l_t = l_s/\left(1 - \overline{\cos\theta}\right)$ where $\overline{\cos\theta}$ is the mean cosine of the scattering angle, which is close to zero, and V is the neutron velocity. In uranium and graphite we have approximately $(l_t)_{gr} \approx (l_t)_U \approx 2.5\,\text{cm}$; in water l_t is much less: $(l_t)_{H_2O} \approx 0.8\,\text{cm}$. Therefore the difference in the diffusion coefficients, $(l_t)_{H_2O}/(l_t)_U \approx 1/3$ must be compensated by the difference in the derivatives: $(\partial n/\partial r)_{H_2O}/\partial n/\partial r)_U \approx 3$. It follows for the particular case of the Chernobyl reactor that the neutron density on the graphite boundary is about 1.5 times higher than the neutron density on the uranium surface (Fig. 2). Now assume that the water gets very hot or even starts boiling. Then (in case of boiling) the neutron density on the boundary of the graphite is equal to that on the boundary of uranium, and then the neutron absorption in the graphite drops significantly. That implies that the reproduction factor rises sharply and the reactor blows up. That is what happened at Chernobyl. But the explosion was a thermal one, and not a nuclear explosion, because the resonance absorption due to the Doppler

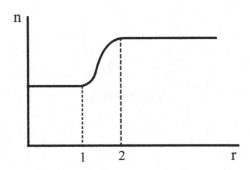

Figure 2:   Neutron density in the cross section of a fuel rod as a function of r, 1 — uranium-water boundary, 2 — water-graphite boundary.

effect increases with temperature and the coefficient becomes negative at very high temperatures (of the order of 2,000° to 3,000°). The positive temperature coefficient was a shortcoming of the design, but the authors did hope that additional systems which they introduced would neutralize its effect. Alas, that did not happen.

In fact, the Chernobyl disaster proceeded like this: an experiment was to be conducted at a power level of 200 MW. At that moment the reactor was operating at 400 MW. Unfortunately, almost all the control rods had been lifted, i.e. they were outside the active core. The operator began to reduce the power, but in view of the positive temperature coefficient the power dropped below 200 MW (Fig. 3). Then the operator began to raise the power to go back up to 200 MW, but the reactor jumped beyond the required value, and hence the explosion occurred. It should be emphasized that the explosion was a steam explosion, not nuclear. The reason for this was that at temperatures of about 3000°C the negative coefficient, arising from resonance absorption and described on pp. 33–35, overwhelms the one from graphite absorption, and the chain reaction stops. Nevertheless, the explosion at the Chernobyl AES was so strong that the 150 tons upper shield of the reactor got turned upside down. The explosion was asymmetric — only half of the reactor exploded. This is why no reactors with large positive temperature coefficients are built anywhere in the world. Moreover, the reactors of nuclear power stations are protected by a hemisphere of metal — a containment capable of withstanding the pressure. This containment has built-in water sprinklers. If pressure rises, these sprinklers squirt

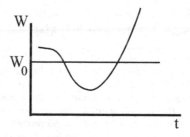

Figure 3: Power W as function of time t, $W_0 = 200\,\text{MW}$.

jets of water inside the containment which produce droplets of water. Water is good at absorbing the most dangerous radioactive substances: iodine and caesium. Therefore even if an accident were to happen, the emission of radioactivity would be rather less than in the Chernobyl plant.

2.9. An Incident With a Nuclear-Powered Submarine, Similar to the "Kursk" Disaster but with a Happy Ending

First, a few words about the very first plants producing electricity by means of nuclear power (nuclear power stations, AES). This is necessary because of erroneous statements in Russian (Soviet) literature. The first AES in the world was built in Oak Ridge in 1951. Its principal creator was Walter H. Zinn,[4] who was also in charge of its launch. Its power was slight — 2 kW, but light bulbs plugged into the net did light up. The second AES was a prototype for the "Nautilus" submarine. And finally, the third (or fourth) AES was "Nautilus" herself, which was launched on 21st January 1954 and which started its first voyage on 17 January 1955. "Nautilus" was created by Admiral Rikover — the father of the American nuclear submarine fleet. The first AES in the USSR, with a power of 5 MW, was launched on 27 June 1954. (It may therefore be counted as third in the world.)

Now I shall tell about the incident as promised. I had a friend — Georgy Alekseevich Gladkov, who was the main designer of reactors on all series of submarines. He also carried out their running tests, took part in long-distance underwater excursions and was granted the award Hero of Socialist Labour.[5] (Alas, he is no more). We

---

[4]W.H. Zinn was the first director of the Argonne National Laboratory from July 1946.

[5]I met him in interesting circumstances. He had designed the first Soviet nuclear submarine and was to report on the project to the PGU Scientific-Technical Council. I was the scientific expert on this project, I had to work out an official assessment and to present it at the same session. Since this was all of crucial importance, we examined every detail of the project. My assessment was positive and in the process of carrying out the job, Goga and I became friends — particularly when we found out that he used to go hiking in the mountains, just as I did. Goga was a very modest man. We were once traveling together on Lake Baikal. There was no cabin space for us and we put up a tent on the

used to call him Goga. Here is what he told me. We had produced a series of submarines with nuclear propulsion which were capable of executing a full reverse. A full reverse is a maneuver in which a submerged submarine proceeds at full speed, suddenly stops, and moves backwards at almost the same speed. It is clear that the ability to execute such a maneuver is of great importance for a submarine.

Well, the running tests of this series were under way. Goga is in the wheel-house standing next to the captain, behind them two admirals on a little sofa. The captain issued the command: "Full reverse!" The submarine immediately experienced a strong shock, it assumed an angle of 42 degrees incline relative to the surface but its speed barely reduced at all. Until that moment the submarine was travelling at a depth of around 100 meters, and the bottom of the sea was at about 300 meters. The captain realized that at that speed and in this position, the submarine would hit the bottom within 30 seconds and that would be the end of it. He did not lose control and immediately gave the command to surface. The submarine popped out of the water like a cork. Once everything had calmed down and the submarine had stopped, the captain called the commanders of all services to report on what had happened in their own domain. The first to report was the head of the engine room, who was in charge of the transformers filled with oil. He said that if the angle of incline had reached 45 degrees, the oil would have poured out onto the floor, the floor was scorching hot, a fire would have started. But the incline was only 42 degrees. Lucky! The next to report was the commander of torpedoes, who said that one of the torpedoes had torn off its fastenings and had started off ahead, but it did not reach the unit that launched its explosion and stopped. Lucky again! Once everyone had been questioned, Goga turned to the admirals. They sat there, pale as pale can be. He went up to them and said: "The plan for testing the full reverse demands that the test be repeated twice. Please give your command about when to carry out the next

---

deck. It started to rain, however, water was running all over the deck and we all found ourselves in a puddle. I told Goga: Go and see the captain, show him your certificate of Laureate of the Lenin Prize. Goga refused. Then I took that certificate to the captain myself, and we were given a cabin for two. Goga refused to go into the cabin, and so did I.

test". The answer was couched in terms of foul language at a level he had never heard before.

2.10. How I Became a Theoretical Physicist

I came back to Moscow in June 1943 (I had been evacuated away from Moscow). I was 17 and had completed nine years of schooling. I had to decide urgently what to do. There was no point in going back to school: I would turn 18 the following year and then I would be drafted into the armed forces. When such boys as myself, inexperienced and ill-prepared for life, were sent to the front line, they would as a rule be done for in the very first attack. (I knew this because in 1942 I had worked in a hospital for 6 months.) At that time (in summer 1943) several Technical Universities in Moscow started preparatory courses: these took on youngsters who had completed their 9th year of education. They would work through the syllabus of the 10th school year, then sit exams, and on passing they would be accepted at that particular establishment as first-year students. Students in higher education had "bronia", i.e. their call-up to the armed forces was postponed: the government had realized as early as 1943 that the country would need engineers after the war. The university did not offer any preparatory courses, those at the Energetics and Aviation institutes were full and did not accept any more candidates. The only establishment which still had places was the Moscow Electro-Mechanical Institute for Transport Engineers (MEMIIT). I became a student there and this is where I completed my first year of higher education in 1944.

In the summer of 1944, first year students of MEMIIT were sent for one month for logging in the town of Konakovo near the Moscow-Volga canal. Four of us were given a small room in a village house. We were prepared for what was in store for us and had brought some pyrethrum — a powder against bedbugs — and used it on the floor of the room and also on its walls up the the height of 1 meter. When we woke up during the night and lit a torch, we saw that above that line the whole wall was in motion — bedbugs were walking up and down it all over. Then we discovered that there were no toilets whatsoever — none in the house, none in the yard — only 100 km from Moscow, we had found the Middle Ages. The eating place (where we were fed twice a day) was situated at a distance of 3–4 km from the village, and the forest where we worked was 3–4 km away in another direction. This meant that in addition to working we had to walk for about 20 kilometers every day. If one did "fulfil the norm", that is, if one managed to complete the set amount of work ("the norm"), one would get the full food ration (the norm) every day: 600 grams of bread to last the whole day, plus in the morning some sort of porridge, and in the evenings first some soup and then some more porridge with a bit of meat or fish. If one could not complete the norm, the bread ration was reduced to 200 grams per day; if one managed to work more than the norm, one would be allowed a "Stakhanovskie norm": the bread ration would rise by 400 grammes and one would get a second helping of evening porridge. We were logging trees in the forest and it was not possible to fulfil the norm. Therefore we adopted the following strategy: one day we would hand over to the foreman only 20–30 per cent of the norm and hide the rest — it did not matter by how much one had underfulfilled the norm — but on the following day we would "overfulfil the norm" and get the "Stakhanovskie" food ration.

I was not really happy at the prospect of working on the railways, I wanted to do physics. I decided to join the Physics faculty of Moscow State University (MSU) as an external student. I needed permission from MEMIIT for this, and this was granted without difficulty because I had a good reputation there. Even more — they granted me the right to attend lectures on a "free regime", instead of a compulsory one. This was extremely rare. In autumn 1944, I became an external student of the MSU Physics faculty and was successful at the winter examination session: I got full marks in every paper. In March–April the Physics faculty announced that they were inviting applications for a special group: students from any other institution could be accepted without any additional examinations, and also if one was accepted to this group, one's original institution was obliged to let this student go, and he was also excused army service. It is now clear that this was a selection of personnel for the atom project, and this was happening before it was announced that the atom bomb had been exploded. I tried to join that special group but I was refused in spite of having an advantage over all other students — I had passed the first part of the university's physics examination. Another person who was rejected was David Abramovich Kirzhnits, a future Corresponding Member of the Academy of Sciences, who had a recommendation from Landau. He got this recommendation in an interesting way. Kirzhnits was a student at the Moscow Aviation Institute (MAI). They had a physics teacher who noted his extraordinary gifts. She was acquainted with Landau, she told him about this gifted student and Landau invited him for a talk. After talking to him, Landau said: I shall write a letter of recommendation to the Dean of the Physics Faculty, Predvoditelev. He took a sheet of paper and a pen — and then mused: I cannot write "Dorogoi (Dear in Russian) Aleksandr Savvich — he said — I do not hold him dear at all. Esteemed ... — I do not esteem him." He thought for a little longer and exclaimed: "I shall write DEAR in English — dear in English does not mean anything at all". In spring I had to sit examinations at both MEMIIT and the Physics faculty. I decided to get satisfactory marks at MEMIIT, so I would get only 3 out of 5 for all railway disciplines, this would make it easier for me to leave the institution. We had a subject called "Fuel, water and lubricants". I practically never paid any attention to it, I just spent a few hours immediately before the examination leafing through the manual. At the examination, I was asked the question: "We have a mixture of water and kerosene. How can we separate the one from the other?" I could not force myself to say anything apart from: "let them stand for a while, and then you will have kerosene on top and water at the bottom". I got top marks.

I got top marks in all subjects at the spring session at the Physics faculty (as an external student) and its professors Gradshtein and Modenov gave me very good recommendations. By autumn I resumed my efforts to become a full student rather than a part-time one. I did not say that I was a MEMIIT student, I only said that I wanted to move from external studies to full-time. I had to go and see the same Assistant Dean of the Physics Faculty, Georgy Petrovich Z, who had rejected me when I had tried to join the special group. The result was the same, and his argument was that a move from external to full student was not allowed. I then went to the MSU Prorector, Professor Spitsin. He said the same thing, but showed sympathy and added: "But do complain, one can complain infinitely". "Complain to whom?" — I asked. "To the Ministry of Education" — answered the Prorector. So I went to the Ministry of Education and saw the Head of Administration for Universities. And there I had luck! The head of this department — Zatsepin (alas, I do not know his first name and patronymic) turned out to be a good man. He said: "You want to move from external to full? Well, just do it". "But" — I asked — "will you sign an official paper saying that you do not object?" "Of course" he said and there and then he dictated the text to a typist. I took this letter to the Physics faculty and was lucky again: Georgy Petrovich was on holiday and another man was standing in for him. Looking at this man with perfectly honest eyes, I said: "Georgy Petrovich promised that if I had a letter from the Ministry of Education saying that the Ministry did not object to my moving from the external Physics department to the regular one, he would enrol me in the regular one. Here is the letter." He answered: "I shall issue the order that this be done". When Georgy Petrovich returned from holiday, he could no longer do anything about it. But I still had to leave MEMIIT and I had to leave without any scandal — otherwise they would have sent my papers straight to the military call-up office. I succeeded in achieving this in 7 visits to the Dean and director of MEMIIT, the railway general Fedorenko.

This is how I became a physicist. But much more time had to pass before I could begin to hope that I might become a theoretical physicist. I hesitated for a long time, but then I took the jump and began to pass the Landau theoretical minimum. The first examinations — the entry exam in mathematics, mechanics, field theory and the first part of statistical physics were relatively easy for me. But I got stuck when studying quantum mechanics: I could not understand some of the questions, they had to be studied by reading original articles, the Landau-Lifshits book "Quantum Mechanics" did not yet exist at the time.

In their fourth year, students were sorted into different departments. I applied to the department of theoretical physics and was accepted. However, a new order sent me to another department — "Structure of Matter". This was a code name, the department was in fact that of nuclear physics. At the time I was very upset by the decision and attempted to have it cancelled, but then I understood that in fact I had been lucky again. The thing is that this particular department allowed the student to choose as supervisor for his diploma dissertation any physicist who was taking part in the atom project. All other departments demanded that the supervisor be teaching in the Physics faculty. In the "Structure of Matter" department, students preferred to find their supervisor and to come to an arrangement with him independently. Kirzhnits and I wanted to choose someone from the Landau school. A student in his last year — B.V. Medvedev — gave Kirzhnits two telephone numbers — of Pomeranchuk and Kompaneyets. We decided that Kirzhnits would telephone first and if he succeeded in his request, I would then call the other number. Kirzhnits started calling Pomeranchuk's number, he called several times but did not have any luck: Pomeranchuk was never there. Then he rang Kompaneyets, who readily acceded to his request. This meant that I had to contact Pomeranchuk. I made many calls and each time a man's voice would answer "No Pomeranchuk here!" Much later, when I was already working at ITEP, I understood what was the matter. The telephone whose number I had been given was located in the hall, not in Pomeranchuk's office. This telephone was manned by a soldier. At last, a miracle happened when I rang — the same voice called out to Pomeranchuk and asked him to the telephone. Pomeranchuk must have been walking past, or maybe he had been standing in the hall talking to somebody. This was another piece of luck for me! I said that I was a student at the university, that I had successfully passed the Landau theoretical minimum examinations in three subjects and that I wanted to do my diploma work under him. The fact that I had passed three Landau minimum examinations was sufficient in Pomeranchuk's eyes to take me on for diploma work: at that time few people had been successful at the theoretical minimum — only 10 people in all — and he invited me to his home for a final discussion. I went. On that day there was a severe frost. I had no fur coat or indeed any coat at all, I was wearing a short summer jacket which my mother had lined with squirrel fur taken from an old coat of my grandmother's. For some reason this fact made a strong impression on Pomeranchuk. This summer jacket decided my fate — Pomeranchuk told me so himself at a later time — he took me on for my diploma. Pomeranchuk helped me a great deal in preparing for the quantum mechanics examination: he gave me galley-proofs of those chapters of the book "Quantum Mechanics" which explained the themes which I could not understand. After this I passed quantum mechanics and several more subjects. I started on the diploma work — Pomeranchuk had given me its subject. However, I was not sure that I was able to become a theoretical physicist. Rather to the contrary, I felt that I did not even have a hope. As one of the heroes in Sinclair's book "Arrowsmith" says: "not everybody working in science is a scientist — only very few are". This is even more applicable to theoretical physics: one may pass examinations and solve problems, but creative work is an entirely different matter.

The first time I could hope that I would be able to become a theoretical physicist happened when I was working on the problem set me by Pomeranchuk. I remember this moment clearly — the star moment of my life.

Pomeranchuk asked me to calculate the polarization of slow (resonance) neutrons in scattering on nuclei. This polarization arises from the interference of nuclear scattering with the Coulomb scattering of the neutron magnetic moment (a relativistic effect). The amplitude of the neutron magnetic moment interaction with the Coulomb field is purely imaginary. Therefore the interference arises only if the nuclear amplitude has an imaginary part. A similar problem had been solved previously by Schwinger, who considered neutron scattering at high energy, where the nuclear scattering is diffractive and its amplitude is purely imaginary. In Schwinger's case, the momentum transfer is much greater than the inverse size of the atom, and the interaction of the neutron magnetic moment with the atomic electrons could be neglected. Pomeranchuk advised me to look at neutron scattering at low energy, in the resonance region, where the nuclear amplitude also has an imaginary part. However, in this case scattering takes place with a momentum transfer of the order of the inverse size of the atom, and the interaction of the magnetic moment of the neutron with the atomic electrons is significant. I tried to find a region where Schwinger's method could be used, but the results were not convincing, and I got depressed. I was sitting in the reading room of the MSU library and — I clearly remember this instant — it suddenly occurred to me that I could do the calculation exactly, not applying any approximations: I only had to take account of the atomic form factor, which I could do by describing the atomic electron cloud by the Thomas-Fermi method. That was my own idea — Schwinger did not have it! At that moment I had hope for the first time that I could become a theoretical physicist.

2.11. How I Prevented an Accident at the ITEP Reactor

This happened around 1955. The ITEP heavy-water research reactor was to be reconstructed: instead of working on natural uranium, the reactor would be working on 2% enriched uranium. The solid cylindrical uranium slabs were to be replaced by ring-shaped slabs, and some construction changes were to be made. As a result, the power of the reactor would increase four-fold, and the flux of thermal neutrons would increase by nearly an order of magnitude. I was carrying out the physical calculations for the reactor. It was the first reactor to be commissioned for which I had full responsibility for all physical calculations. (Before this I had done independent calculations only for possible future reactors, which in fact did not get built. The responsibility for the calculations had remained with A.D. Galanin, I just did the work as instructed.) Well, the day of the physical start-up of the reactor came. The head of the start-up operation S.Ya. Nikitin invited me to be present at this experiment. The physical start-up of a heavy-water reactor happens in the following manner: uranium rods are inserted into the reactor in the absence of the moderator — heavy water. Since there is no moderator, there is no chain reaction, there also is no flux of neutrons. Then heavy water is gradually added. Once the heavy water reaches a certain level, the reactor comes up to its critical state, the chain reaction starts, the reactor "is running". The critical level of heavy water, which is predicted by physical calculations, is the main parameter for the continued functioning of the reactor. When its experimental value coincides with the theoretical prediction, this indicates that the theory is sufficiently reliable and the operation of the reactor may be carried out as predicted. If there is a contradiction between theory and experiment, unexpected things can happen. Before launching the reactor Nikitin asked me what the predicted critical level was and how accurate it was. I said that the value of the level was 150 centimeters and that the error in this value must not go beyond 5 centimeters. They started to pour in the heavy water. At the same time the neutron flux N was being measured in several places (an artificial source of neutrons had been placed on the bottom of the reactor) and a graph of $1/N$ as a function of the heavy water level was plotted. Obviously, at criticality $(N \to \infty)$ the curve $1/N$ had to cross the abscissa. The 10 cm level was reached, then 5 cm below the expected critical level — the curve $1/N$ did not tend to the point that I had predicted. Nikitin tried to comfort me: "It does happen that at the last moment the curve turns". We reached the predicted critical level — the reactor does not run. We go further, going 5–8 cm higher than that — it does not run. One could clearly read on the faces of the experimentalists and engineers present their thoughts: "The calculations for the first reactor were done by Galanin and Pomeranchuk, and now look what happens when such a responsible job is entrusted to youngsters". Another 5 cm of heavy water was added — the reactor still did not run. At that moment Nikitin ordered the launch to be stopped and reported to A.I. Ioffe on what was happening. Ioffe was not at all pleased — this was great trouble for him. He may well have had the same fleeting thought as the experimentalists had had. He did however postpone all further work on the launch of the reactor until the following day and told me: "Check all your calculations and report the results to me tomorrow". All through the evening I checked my results, together with Rudik who came to help me, but no mistake was found. I could not sleep at all that night, but I gathered my courage together in the morning, went to Ioffe and said: "I cannot see any mistakes in the theoretical calculations. Such a large discrepancy between theory and experiment ought not to be present". At this point Ioffe gave in and ordered: "Do not carry out the launch, remove the water, have the engineers look for the mistake in their own work". Two days later I was visited by B.A. Medzhibovski, an engineer who dealt with the reactor control system and who had nothing to do with its assembly, and asked: "If the uranium rods are hung not as they ought to be according to the design but 20 cm higher, what will the critical level be then?" I made a quick calculation and said: "Right at the point to which the $1/N$ curve was tending". Medzhibovski explained that on studying the blueprints he had found a place where the rods might have been affixed: this place was very similar to the correct one, but situated 20 cm higher. He had immediately gone to Nikitin to tell him about his guess. Nikitin had called the head of assembly, the senior mechanic A.P. Shilov. Shilov immediately started shouting: "Rubbish! This can't be! Never!" Then Nikitin ordered the top lid of the reactor to be removed. He said that the next day he would measure himself the placement of the rods and asked me to be present for this. When I arrived, the lid of the reactor had been removed, Nikitin was standing on top of the reactor wearing spectacles with dark glasses, gloves and an overall. It is possible that he was wearing something under the overall. It must be said that it is not really safe to be above a reactor with its lid off. Although the reactor had not been launched, there had been something of a flux of neutrons, and this means that there was also radiation. Therefore all those present were told to keep away from the reactor at a good distance. Nikitin took a long dowel, plunged it into the reactor, marked something on it, then took it out and used a tape to measure its length up to the mark that he had made. This he did in several places in the reactor. After this he announced: "The rods are wrongly sited, they are 20 cm higher than they should be. I shall report to A.I. Ioffe". The reactor had to be reassembled. If the reactor had been launched with its installation faulty in this way, the top ends of the uranium rods would have been above the level of the moderator. This would have strongly speeded up the radiation, due to fast neutrons, and would have led to extremely unwelcome consequences — the nearest dwellings were only 100–200 meters away and people were living in them.

2.12. Nuclear Reactors and Politics

I shall begin with an episode relating to B. Pontecorvo's arrival in the USSR. At the end of the 1940s Pontecorvo was living in England. Early in 1950 he went with his family to Finland, ostensibly for a holiday. The Soviet steamer "Beloostrov" was waiting there for them and took them to the USSR. The operation of their departure from Finland was carried out illegally and it was only later, when Pontecorvo had gone, that Western intelligence established that this was the way he had disappeared. There was no mention in our press about his arrival, and I for one got to know about it much later, on reading the American journal *Science News Letters*. Once in the USSR, Pontecorvo lived and worked in Dubna. He was prohibited to leave Dubna until approximately 1955 — he lived there in a sort of exile. It was prohibited to mention his name. Pomeranchuk, who used to travel to Dubna frequently in those days, said many times on return from there that he had discussed some question or other with "the professor", and that "the professor" had said this and that. The "professor" in question was Pontecorvo, but Pomeranchuk never mentioned his name. This continued until 1954.

At some time in 1950 Galanin was unexpectedly called to the Kremlin. Such a call was very unusual indeed: people were called to all sorts of places, but never to the Kremlin. Since Galanin was working with reactors, it was obvious that this call was in some way connected with reactor activity. Usually Galanin would discuss all problems to do with reactors with Rudik and me: we too did calculations on reactors — it would have been impossible to work otherwise. But on this occasion, Galanin came back from the Kremlin — and did not say a word. In those days, theorists obeyed the rule introduced by Pomeranchuk: no questions. As he used to say: "Those who need

to know will hear it from me, I'll tell them myself". Therefore we did
not ask questions. Galanin kept this silence for a long time, in fact
for several years, but eventually he did break it. It turned out that
he had been called to the Kremlin for Pontecorvo's interrogation. A
group of physicists had been called together and it was suggested
that they ask Pontecorvo questions about what he knew concerning
the atom problem. But Pontecorvo only knew general principles, and
those assembled were mostly interested in technical details: for instance about the construction of a reactor's uranium slabs, about the
technology of various specific processes etc., but Pontecorvo did not
know anything about this and could not give any useful information
on that occasion.

Pontecorvo's contacts with physicists were severely limited. Pontecorvo was not allowed to publish any scientific articles — his name
completely disappeared from science for five years. Nonetheless, he
did not change his communist views. Somewhat later, in 1956, we
both attended the conference on elementary particles in Yerevan and
shared a hotel room. Pontecorvo had just returned from a visit to
China which he had made as a member of the Soviet delegation. One
evening, when we were already in bed, he began to tell me about his
impressions. He was full of enthusiasm about what he had seen: how
wonderful the communes were, how strong the enthusiasm of the
people in building communism etc. I could not refrain myself and
remarked: "Bruno Maximovich! If one observes a country from outside and if one visits it as a guest for a brief time, one can be badly
mistaken". Bruno Maximovich broke off the conversation, saying:
"Let us sleep". He never forgave me for this remark: we had been on
very good terms before, but our relationship was never reinstated.
The conflict with China struck a year or two after this conversation.

As far as our relationship with China goes, Pomeranchuk's foresight was very much better. As early as at the beginning of the 1950s,
at the time of the song "Moscow–Peking", he was predicting for the
future the most serious conflicts and possibly even a war with China.
It is true that this prediction had been made in Orwell's book "1984"
which had been published in 1949. But at the time we did not know
that the book existed.

As Dubna has already been mentioned, I will tell a story which was told as an entirely true story: about the way in which the International Joint Institute for Nuclear Research in Dubna was organised. At that time, it was called the Hydrotechnical Laboratory (HTL) — presumably for the reason that it stood on the Volga river. But it had nothing at all to do with hydrotechnology. This institute was created at the suggestion of I.V. Kurchatov to do research in nuclear and elementary particle physics. In actual fact, the research undertaken in Dubna was in no way related to nuclear weapons. (Though for many years the powers-that-be were persuaded that the opposite was true.) At the time of making the decision to create the Institute, the natural question of its siting was discussed. A special commission was set up to study this question. Beria called a meeting at which the commission presented its recommendations and suggested three possible sites for the future institute. Beria listened, then asked for a map, pointed a finger at a place for the future Dubna (this was not one of those recommended by the commission) and said:

– We'll build it here.

– But — somebody timidly objected — this is an area of swamps, the ground conditions are not suitable for accelerators.

– We'll drain them.

– But there are no roads.

– We'll build them.

– But there are few villages there, it will be difficult to find laborers.

– We'll find them, said Beria.

And he was proved right. This area was surrounded by labor camps, this is why Beria had chosen it. As late as 1955, when I first visited Dubna, the camps stretched along the road, there were guards everywhere and one had to tell them "We are going to see Mikhail Grigorievich". (This was M.G. Meshcheriakov, director of HTL.)

Nuclear reactors in China were built on the basis of Soviet projects and mostly the building work was carried out by our technical specialists — China did not have any of these at the time. The head of the Chinese nuclear program, Tsian, decided to start it with the building of a heavy water nuclear reactor. ITEP was appointed to

work out the design of such a reactor, also to send to China the specialists needed to build and launch it. I was given the task of doing the
physical calculations for this reactor. Three Chinese physicists came
to us in ITEP in order to be taught by me. One of them was Peng, a
theorist who had been working with W. Heitler in the 1930s. In the
1950s he already was an academician and his function was mostly
to be an official representative. Another was obviously the group
commissar, he was not interested in science, he had other duties.
Only the third, a young man called Huan, turned out to be capable
and willing, so that he succeeded in mastering this science in a short
time. The research reactor in China was built very quickly and was
launched in 1959. (This reactor worked for about 30 years.) At the
same time as building research reactors, Soviet assistance was given
in the construction of military reactors for the production of plutonium and chemical plants for its isolation. An order came from above
to furnish China with the most modern projects that the USSR had
at its disposal. Physicists and engineers who were to carry out this
task, understanding the political situation better than the people at
the very top, attempted to send older projects over. However, Zadi-
kian, USSR Adviser on nuclear matters to the Chinese Government,
caught them at it and reported on them. As a result, the USSR sent
to China its most advanced technology. The breakdown of relations
with China came soon after.

2.13. Designing and Building a Nuclear Power Station in Czechoslovakia

I shall relate another story connected with nuclear energy. It is
interesting in that it sheds light on the hidden mechanisms which
were at work in this field, in particular in its international aspect.

It is well known that Czechoslovakia is very poor in energy resources. All the hydro-resources — very limited ones — were long since
fully used. It has only small deposits of lignite. But it does have
uranium ore. (Immediately after the war, these uranium mines were
put under the control of the Red Army and all the uranium mined
was sent to the USSR.) Therefore the government of Czechoslovakia

Figure 4:   General view of AES A-1.

decided to develop nuclear energy in the country and asked the Soviet
Union for help. In 1957 it sent a governmental delegation to Moscow
in order to sign agreements on building nuclear power stations in the
People's Republic of Czechoslovakia with our assistance.

The Soviet side proposed several projects of nuclear power stations designed by the Institute of Atomic Energy (which worked on
enriched uranium) and an ITEP project for a heavy water reactor
working on natural uranium. Let me remind you that in 1957 — under Kurchatov — monopolism was not quite as strong, competition
was still allowed, so that our Institute's project had more or less the
same authority as those presented by the Institute of Atomic Energy
(IAE, now the Kurchatov Institute).

The Czechs chose the ITEP project. They had the following considerations. They did have their own uranium, but they did not
have any diffusion factories for its enrichment. Therefore if they

were to build nuclear power stations working on enriched uranium, they would become entirely dependent on the Soviet Union for their energy. If they had power stations working on natural uranium, they were intending to ensure (if not straight away, then at some later date) that the uranium from their own mines would go directly to their own power stations. Of course, the power station that we were proposing was more complex both to build and to run. But the Czechs did not worry about this — the level of industry in Czechoslovakia was sufficiently high. Moreover — as the Czechs themselves told me later — they had far-reaching plans for the future: they meant to develop the technology and industry needed for chain-building such power stations and to present these to the world markets, where small and under-developed countries would buy them. They intended thus to create for Czechoslovakia an energy and economy independent from the Soviet Union. This line of thought was firmly adhered to by all governments of Czechoslovakia until 1968 — were they of the orthodox-communist persuasion under Zápotocký and Nowotny or of that under Dubček.

The scientific management of the project was carried out by ITEP, the scientific manager was A.I. Alikhanov. I was managing the physical calculations for the reactor. (In those times, the word "manage" in ITEP did not have the meaning which it usually has nowadays. Managing the physical calculations meant that a man had to calculate everything concerning the physics of the reactor, or at least to carry out a detailed check of the calculations done by others.)

One day I had a knock at the door of my office at ITEP. It was a man whom I did not know. He introduced himself: "I am the Chairman of the Nuclear Commission of Czechoslovakia, Avgustin Shevčik" (that meant — the Minister). "I am an engineer, he said, but I do not know anything about nuclear physics or nuclear reactors. I would like to learn about these".

I answered: "This will be a lecture course and you will be given homework: some problems which you will have to solve yourself". He said: "I agree".

I gave a lecture course to Shevčik and he (a Minister!) did solve the problems which I set him as homework. Shevčik remained at the head of the Atomic Commission under Zápotocký, Nowotny and Dubček. Under the government of Štrougal, after Soviet troops had entered Czechoslovakia, he was dismissed from this post and became just a consultant at a power station. But we remained friends and each time I came to Prague I would telephone him and we would spend the evening in some café — he knew good cafés in Prague.

The power station was duly built in Jaslovske Bohunice (Slovakia), it was called AES A-1. At the beginning, the launch of the plant was planned for 1965–66, but the work was slow, completion dates were pushed further and further, and at last it was decided to formulate a final version of the launch program for early 1968, which included sending a Soviet delegation to Czechoslovakia. But here the Prague Spring events took place and the Soviet leadership decided that it was imperative to wait a while. They waited until the moment when our army was sent to Czechoslovakia and a new, pro-Soviet government came to power, headed by Štrougal. At that moment attitudes changed radically, it was decided to give great prominence to the launch of the power station as proof of the friendship between the Soviet Union and Czechoslovakia and as proof of the Older Brother helping the Younger, once the latter had returned to the correct path.

To characterize the level of Soviet diplomats who directed this process, I'll give the following example. The economics attaché of the Soviet embassy in Prague sometimes came to deliver lectures, the Soviet specialists who worked on the AES. On one occasion he said: "We published a book on the Czechoslovak language ...".

A Soviet delegation was to travel to Czechoslovakia in November 1968 and there was a firm order that there was no question of any failure in the work. This helped me to travel abroad for the very first time — until then I was not allowed beyond the frontier. N.A. Burgov, who was in charge of the launch, declared that he could not guarantee success without me, the man responsible for the reactor's calculations. Before traveling, our delegation had to undergo a briefing at the Committee for Atomic Energy — that was the general rule — first in the department of nuclear power stations, and then in the regime department, that is the classified department. This latter

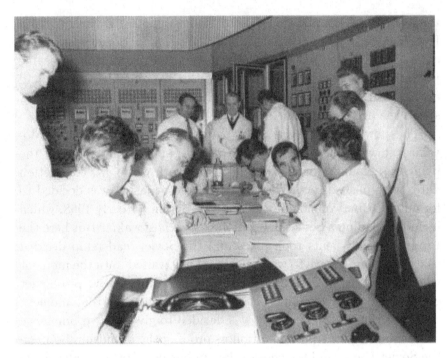

Figure 5:   The A-1 reactor critical experiment. Sitting (right to left): V.F. Belikn, B.L. Ioffe, B.I. Ilyichev; Standing B.A. Medzhibovsky.

briefing turned out to be quite out of the ordinary. The assistant head of the department told us: "We cannot give you any instructions, we do not understand ourselves what is happening or how you ought to behave. We put our hopes in you. Just act according to circumstances".

The talks were held at the Skoda factory in Plzeň. The general atmosphere — if truth be told — gave little ground for joy. The very people with whom we had done much successful work before this and with whom we were in stable friendly terms when they came to Moscow and also when some of us visited Czechoslovakia, now sat across the table from us with stony faces and all wore the Czechoslovak flag in their buttonholes. Even coffee was served only to the Czechs during these talks. Later I heard the explanation that this behaviour of our partners was due to their fear: they feared their party organisation and their trade union organisation, which were very strong at Skoda and which at the time were maintaining a very strong position against all Russians. The more so because the situation in the country was most oppressive: in the streets, on the tarmac we could see gigantic slogans "Ivan, go home!"; on Wenceslas Square in Prague, where our tanks had shot at the Parliament and at the crowd, young men were standing as a guard of honour, holding candles; protest demonstrations were held in factories. Even if I did not approve of the invasion of Czechoslovakia and this had been the heaviest shock for me, despite not hiding my feelings about this — I still felt sharply that I carried part of the blame.

However, from the business point of view the talks were fully successful. The program of the launch was worked out and signed — but then the following happened: The majority of the Czech specialists who had taken part in our work — engineers and even mid-ranking technical personnel — were people of liberal views, supporters of Dubček. Therefore after the accession to power of ortho-dox Communists, all of them were subjected to repression: some lost their jobs altogether, some were demoted, some were expelled from the Party and so on. A whole layer of society (consisting of the most qualified specialists) was removed. Even this was not enough. The new people who replaced them also appeared insufficiently secure from the political point of view, so this layer too was removed. As a result, the level of professional competence of staff was drastically lowered.

The Central Committee of the CPSU and the Government of Cze-choslovakia took the decision to stress the special importance of the power station's launch: it was to be a demonstration of the assistance that the USSR was rendering Czechoslovakia. The station was still not complete, but it received visits of a whole series of high-ranking visitors from both countries: ministers, a vice-chairman of the Soviet Council of Ministers, even Štrougal himself. The immediate control of the work from the Soviet side was entrusted to Petrosiants, the Chairman of the State Committee for Atomic Energy. The date of the launch was set for the end of 1972, and by autumn of 1972 more than one hundred Soviet specialists were already working at the sta-tion. Petrosiants arrived there himself and officially announced the exact date of the launch. It seems that this timing had something to do with some special date or some special event in Moscow, at which he had to make an official report. Work did carry on, but it was clear that the reactor would not be launched on the date named by Petrosiants. Therefore some tricks became necessary. One of these tricks was demonstrated at the visit of some important member of the government of Czechoslovakia. He knew that at the moment of launching, heavy water would be poured into the reactor. So he was indeed shown heavy water being poured by a workman into the funnel of a tube leading into the reactor. (I even have a photograph of this event.) In actual fact, it was as yet too early for heavy water to go into the reactor. Therefore the tap leading inside the reactor was closed and the heavy water ran down through a pipe to the floor below, where another workman collected it into a bucket.

At last all preparatory work had been completed. But it turned out that for technical reasons the reactor was hot. The physical launch of the reactor and the whole large program of experiments, planned to take a full month, could only be carried out on a cold reactor, this is the only way to check all the parameters on which the calculations are based. And consequently these must be known in order to calculate the operation of the reactor's work at power. Therefore we had to wait for the reactor to cool down before starting the physical launch. The reactor was huge — 150 tonnes — and it needed 3 days to cool down. But the date announced by Petrosiants was practically reached, he could not wait. He demanded that the reactor be launched straight away, he shouted and made threats. The manager of the launch and the head engineer remained steadfast for 2 days, they understood that if the reactor was launched while hot this would make all experiments impossible and also that the exploitation of the reactor would always be carried out blindfold. At the end of the second day they submitted to Petrosiants' threats and announced that the launch would take place on the following day, although the reactor would not have cooled down completely. In the morning (we started work at 6 am and the hotel in which we lived was 30 km away) I arrived at the power station, sat down at a table in the control room and asked the engineers to measure the temperature in the reactor in the places where this was possible, in order to modify my calculations: these had been done for a cold reactor. Petrosiants came up to me and asked: "What is your prediction concerning the critical level?" I told him — "I cannot tell you right now, the reactor is hot and the heat distribution is uneven. I have requested temperature data in order to make corrections to my calculations" — "Yes, that's what I thought, said Petrosiants, you can't say anything" — and moved away. Some time later I was handed the temperature data and I started on my corrections. Petrosiants turned up again and asked — "Well, where is the prediction?" — "I shall give it to you in half an hour", I replied. "I know what you will do — said Petrosiants — you will give a prediction with an error sooo big" — and he showed with his hands the sort of gesture a fisherman makes when he tells about a fish he has caught. Half an hour later I went up to Petrosiants, I told him my data, the error was of the order of three per cent, and I asked him: "What do you say, is this indeed sooo big an error?" He was forced to admit that it was not. The launch was carried out and the critical level coincided with my prediction.

The reactor had been launched, Petrosiants made his victorious report to Moscow, there followed much trumpeting in the press, the station was brought up to full capacity and it worked successfully for several years. But this situation did not suit our government. It wanted to keep the key to the energy of Czechoslovakia in its own pocket. Therefore it started pressurising the Czech government to have all subsequent power stations built for working on enriched uranium, of the WWER type. And the Czech side gave in. At the same time, using two not very substantial circumstances, it was decided to close down and disassemble the A-1 power station. Until the "Velvet Revolution", the whole of nuclear electricity in Czechoslovakia was produced by WWER power stations. Since the "Velvet Revolution", the power stations in the Czech Republic and in Slovakia have been built to Western designs. Nowadays power stations with heavy water reactors are being built in a number of countries, but Russia has dropped out of this activity.

To finish the discussion of nuclear power stations, I would like to move to the problem of their safety — after the Chernobyl disaster this is the No. 1 subject in any discussion of nuclear power stations. In my opinion, the main and irremediable fault of power stations working on reactors of the RBMK type ("Chernobyl type") lies in the positive and high temperature and steam coefficients of reactivity. This means that the reactor as a physical system reacts to the rise of temperature or of the volume of steam by increasing its power. And the opposite is true as well: it reduces its power in response to a fall of temperature or a reduction of steam volume. In other words, it is unstable in principle. This is a cardinal defect of the reactor and it is due to the fact that the neutrons are thermalized in graphite while the cooling of the reactor is achieved by water. One cannot eradicate this fault, this is precisely the reason why energy-producing reactors of this type are no longer used anywhere in the world. Positive coefficients of steam and temperature were actually the cause of the Chernobyl disaster (see Sec. 2.8). There were of course other, additional circumstances which contributed to it, but in my opinion this is the main cause. Those who favour RBMK reactors are actually busy removing such circumstances. In my opinion, any safe nuclear reactor at a power station must first of all be stable as a physical system, that is — it must have a negative (and preferably a high) temperature coefficient (and also steam coefficient if the reactor is cooled using water or if it may come to the boil). This is precisely the characteristic of heavy water reactors which work on natural or slightly enriched uranium of the type described above. Unfortunately, all attempts to build nuclear power stations of this type in our country, or at least to carry out a serious comparison between them and WWER and RBMK types have until now encountered an impenetrable wall of resistance — the monopolism that we have already mentioned. In 1974, after the launch of A-1 power station in Czechoslovakia, I wrote an article in which I gave the description of parameters and results of the launch of A-1 in Czechoslovakia. At the end of the article I added a brief chapter in which I compared heavy water as cooled power stations working on natural uranium with WWER and RBMK stations with respect to the expenditure of uranium for each unit of energy produced (not in respect of safety considerations, since it was obvious that the article would definitely not be allowed to be published). The comparison did not favour WWER (RBMK),[6] although for the latter I had taken the design data which did not justify themselves in exploitation. The Atomic Energy Committee in the person of the head of Nuclear Power Stations department prohibited the publication of my article. The official conclusion stated that the article might be published only on condition that the chapter containing the comparison of different reactors be thrown out. All attempts to overcome this prohibition ended in defeat. At last I managed to reach A.P. Aleksandrov (he was at the time President of the Academy of Sciences, Director of the Atomic Energy Committee and Chairman of the Scientific and Technical Council of the Ministry of Medium Machine Building — which means that he was the Head of the atom project). Aleksandrov wrote on the title page on my article: "Everything said in this article is correct, but the fact that we are building WWER (RBMK) is due to entirely different reasons". As I understand it, the reasons Aleksandrov had in mind were that from the technological point of view the RBMK reactors are close to the military ones and only a minimal reorganisation of industry is needed to build them. After this resolution of Aleksandrov's, my article was published in full. Until the Chernobyl disaster this was the one and only article in Russian specialist literature which expressed doubts about the RBMK (WWER) being the best nuclear power stations.

Today, the times of "enlightened monopolism" in our science evoke only feelings of nostalgia.

---

[6]WWER is the acronym for Water-Water Energy-producing Reactor, RBMK for reactor of high power, boiling; they name the same reactor, sometimes one acronym is used, sometimes the other. All nuclear power stations in Russia are equipped with reactors of this type.

Participants in the Soviet Atom Project

———————————— • ————————————

3.1. L.D. Landau [24–27]

3.1.1. *The Landau theoretical minimum*

I shall start by describing how I became a pupil of Landau's. While in my third year of studying Physics at the Moscow State University Faculty of Physics, I realised that I wanted to be a theorist, but I was not sure I was up to it. I had the feeling that David Kirzhnits — another member of my study group — was more gifted and that he could manage it, but one could not tell whether I could. After some deliberation I did apply to the theory study group and was accepted. The teaching of theoretical physics was weak (I could see this even at that time, in 1947), since all high-class theorists — Landau, Tamm, Leontovich — had been pushed out. Instead, the faculty was given over to great specialists in Marxist-Leninist philosophy, who rejected quantum mechanics as well as the theory of relativity. As my fellow-student Gertsen Kopylov wrote in his poem *Evgeny Stromynkin*

> I was present on the day when Lednev[7]
> collected the kahal of professors
> and with a fearless foot threw kicks
> at Einstein, the lion who had lost his strength.

---

[7]Professor of mathematical physics at the Physics Faculty.

L.D. Landau

Well, in summer 1947, gathering all my courage, I went ahead — I telephoned Landau and asked him whether I might present myself to him and start to sit for the theoretical minimum. He told me to come within the next few days. I had relatively little difficulty in passing the entrance examination in mathematics and Landau gave me a typescript of the program for all seven remaining subjects (in fact there were eight of them, the additional being Mathematics II — complex variables, special functions, integral transformations, etc.) At that time only a few of Landau's textbooks had been published: Landau and Piatigorsky *Mechanics*, Landau and Lifshits *Field Theory*, *Theory of the Condensed State* and the first (classical) part of *Statistical Physics*. For all other subjects one had to study using various books and to a considerable extent also original articles. These articles were in English and German — for instance, the quantum mechanics course comprised two long articles (about 100 pages each) by Bethe in *Annalen der Physik*. In other words, it was taken for granted that one had to be competent in both languages in order to try for the theoretical minimum.

This is how the examination was held: The student would telephone Landau to say that he would like to be tested for a specific course (the order of these courses was more or less arbitrary). "All right, do come at ... " and the time was given. The candidate had to leave all his books, notes, etc. in the entrance hall. Landau would take him to a small room on the first floor which contained a round table with several sheets of clean writing paper, a chair, and nothing else. Landau would set the problem and leave the room, but he would return every 15–20 minutes and look over the candidate's shoulder to see what he had written. If Landau remained silent, this was a good sign, but sometimes he would grunt and this was a bad sign. I have no personal experience of what happened in the cases when the candidate failed the examination (I only know that one was allowed to try again). I came close to the danger line only once, when tackling statistical physics. I began to solve the problem using a method which differed from the one Landau expected. Landau came in, looked over my shoulder, grunted and went out. Twenty minutes later he came again, looked and grunted in an even more dissatisfied manner. At this stage Lifshits entered as well on some errand of his own. He too looked at my notes and shouted: "Dau, don't waste time, just throw him out!" But Dau retorted: "Let's give him another 20 minutes". By that time I had worked out my answer, and this answer was correct![8] Dau saw the answer, looked again at my calculations and agreed that I was right. He and Lifshits asked me a few simple questions, and I had passed.

The problems set by Landau were fairly complex, a student needed about a hour to solve each of them (as a rule, each exam consisted of two difficult problems and one that was easier). This meant that one had to practice one's problem-solving a great deal when preparing for the exam. In order to acquire this experience, I was doing all I could to find problems to solve. (It must be said that there were no published collections of problems, in fact those which are now published in Landau's Course could not be found anywhere.) I would turn to Abrikosov (he had passed the Landau minimum before me) about the problems he had been set (but not their solutions!) and I would solve them. After several examinations I discovered that Landau had a fairly restricted set of problems: sometimes he set me the same problems that he had given to Abrikosov. I think Landau understood that those preparing for his examinations were telling each other what problems they had been set, but this did not worry him: just seeing a student's approach to solving a problem was enough for him to assess his ability and knowledge. Here is an example: a problem in macroscopic electrodynamics. A sphere of a dielectric with electric permittivity ε_1 and magnetic permeability μ_1 rotates with an angular frequency ω in a medium of permittivity ε_2 and permeability μ_2, in a constant electric field **E**. The angle between the axis of rotation and the vector **E** is equal to α. Find the electric and magnetic fields inside the sphere and in the medium.

And here is an episode which is characteristic for comparing the level of teaching at MSU with the Landau theoretical minimum. In spring 1948 it was time to sit the university examination in quantum mechanics. The course was given by Blokhintsev, but I did not attend

---

[8] "Dau" was what Landau's close friends called him to his face, and we did behind his back.

his lectures. I was studying quantum mechanics by following the theoretical minimum program and considered my level of knowledge as yet insufficient to present myself for a Landau examination: I needed much more work for this. One day, crossing the MSU courtyard, I met D. Shirkov, who was a student in the Theory Department. "I am on my way to take the quantum mechanics examination before the official exams start. Like to join me?" — "Yes, OK" I said after a minute's consideration. We passed the examination, I got the top grade and Shirkov got the grade below. But I managed to get the top grade with Landau only in September, after another three months of studying for it.

It took me almost two years to pass the whole minimum program. (Over the same two years I also completed two scientific projects under Pomeranchuk.) In June 1949, once I had passed the last examination, Landau included me in the list of his pupils.

About two or three weeks before the tragic car accident on 7 January 1962 which cut short his creative life, Landau wrote out a list of all those who successfully passed the theoretical minimum examinations. Until 1951 Landau examined everybody himself. In the 1950s, however, the number of people wishing to sit the examination increased sharply, and in 1951 Landau decided that he would only see everybody at the first (entrance) mathematics examination. All others would be run by his staff at the institute of Physical Problems: Lifshits, Khalatnikov, Abrikosov and others. Now that many years have passed, we can say clearly which of those who passed the minimum did become significant physicists and which remained at an average level. One can discern a fairly sharp boundary precisely in 1951–52: the number of well-known theorists is much higher before 1952 than after. The thought arises that what was important was not only the content of the program and the choice of problems at the examination — what was important was the role of the examiner. Landau could probably see during the examination who was truly talented and who was not. His pupils were apparently not as successful at it. A great man is unique.

But even Landau sometimes made mistakes. The name of V. Khoziainov does not appear on lists of the theoretical minimum, although he did successfully sit the exam in 1950 (or 1951). This is not an oversight of Landau's. Khoziainov was a Physics faculty student in the same year as myself, though he was somewhat older. In our third year we were distributed among specialties but he did not choose theory, he applied to some other department and was enrolled there. But when several students (including me) were transferred from the Theory Department to the Properties of Matter Department, the management of the Faculty of Physics decided to strengthen the Theory Department. "Strengthening" always also meant a "political strengthening". Khoziainov and another student were by decree transferred to the theory department. Khoziainov was a party member, he may even have been a member of the Physics Faculty Party Committee. This is how he became a theorist. I was extremely surprised later that he had been examined by Landau personally for the minimum award. I heard about this from Landau himself. Landau then added that he intended to take him on as a postgraduate student. I tried to dissuade him, I told him by which route the man had got to be a theorist, I told him that in my opinion Khoziainov was quite a weak physicist, and a fairly dubious personality. All this had no effect on Landau, he had just one answer to all my arguments "But he did pass the minimum examination!".

Some time later (maybe a year-and-a half or two years later) Landau gave me Khoziainov's thesis and asked me to give my opinion on it. The thesis was on particle physics, but it was a formal one (I do not remember its content) and my assessment was rather short of enthusiasm. Landau asked: "Does it contain any nonsense? Does it contradict anything?", — I answered: "No, but there is little content". "Never mind," — said Landau — "this means we can go ahead with his defence". Khoziainov did get his degree and straight away started an impressive career. Within a short time he became Secretary of the Party Committee of the Institute for Physics Problems (IPP). Let me remind you that this was in 1952, right in the middle of the struggle against cosmopolitanism — the anti-Semitic campaign. And we are talking about the Theory Division of the Institute of Physics Problems (IPP) headed by Landau, where the percentage of Jews was well over all admissible limits. In fact, apart from Khoziainov there was only one person whose passport stated "nationality — Russian". This was Abrikosov, and in any case only his passport said that he was a Russian, in fact he was half Jewish. To rectify the situation, the IPP created a second Theory Division, headed by V. A. Fock. Fock intensely disliked the role that he had to play, but apparently he could not refuse it. Being the Party Committee Secretary, Khoziainov undertook energetic measures to replace Landau and to put Fock at the head of the whole Theory Division. (It is however not impossible that he was not the instigator of this move, he may just have followed instructions.) He did not have enough time for this — Stalin died. Soon after, Khoziainov was dismissed from IPP. Landau never mentioned him again.

3.1.2. *Landau's seminar*

Being Landau's pupil did not carry any privileges, it only brought duties, since anyone could engage in scientific discussions with Landau and receive his advice. Only few of those who were successful at the minimum became his postgraduate students. A pupil of Landau's had one right, which was also a duty: to take part in his seminars as a full member. But here too: anyone could take part in these seminars, anyone could ask questions and make comments. The duties of the "full members" consisted in the obligation of regularly presenting review talks at seminars. This happened in turns, in strict alphabetical order. After each seminar, Landau would take the latest issue of the Physical Review (in those days it was not divided into sections) and indicated to the person whose turn had come which articles should be reported on at the following seminar. As a rule, there were 10–15 such articles, and they were from different branches of physics. They mostly dealt with experiments or with both experiments and theory. Sometimes we had brief papers on theory — something like Letters to the Editor. I remember presenting a Letter to the Editor of *Physical Review* by Marshak and Tamor, which gave results of a perturbation theoretical calculation of photoproduction of π mesons on nucleons and the capture of π mesons in hydrogen.

When speaking at the seminar, one had not only to report on the article by presenting its fundamental idea and principal results, one also needed to have a clear understanding of the way in which these results had been obtained, to be present and explain to the audience all the essential formulæ and even the experimental techniques and equipment. Moreover, the main duty for the lecturer was to have his own opinion on the validity of the work on which he was reporting. In other words, when reporting, one assumed responsibility for the work (and the mistakes it contained!) to the same extent as the author of the work. And all this was valid for all articles, devoted to all the varied spheres of physics — from elementary particle physics and nuclear physics to properties of metals and liquids.

Landau had a great affection for alum. He always marked articles about alum in *Physical Review*. Therefore in our group (in ITEP) the word *alum* became the nickname for any subject at Landau's seminars that we found of little interest. (I did however report most conscientiously on articles about alum.) Landau knew each and every subject under discussion well (in spite of his hardly ever reading the articles, he just listened to our presentations), he asked questions which demanded an immediate and concrete answer: generalities like "the author states" were not tolerated. There were always specialists on the subject under discussion and they too asked questions and made comments. In other words, it was definitely not an easy job to report on a *Physical Review* article. Fortunately, one's turn came only about twice a year.

Sometimes Landau felt that the lecturer was presenting the article at an insufficiently professional level. In this case he would interrupt him and would ask him to move to the next item. If this was repeated two or three times during a presentation, Landau would exclaim "You have not done your homework! Alesha, whom do we have next?" (Alesha Abrikosov was the seminar secretary, he was in charge of the list of presenter-lecturers.) In the worst cases, if the same lecturer had been chased off the podium several times, he was ostracised — that is, he was taken off the list of seminar members. Landau would refuse to discuss anything with this person, who would however of course still be able to attend the seminars. I remember two such cases, and one of these involved V.G. Levich, a well-known physicist, who later became a Corresponding Member of the Academy of Sciences. Some other pupils of Landau's were sometimes ostracised too — Berestetsky, Ter-Martyrosian and even Pomeranchuk, but this was done as a punishment for other misdemeanours, not for failures at seminars. I shall talk about this later. The delinquent could be pardoned only after a long period of time — a year or more — and only after one or two of the most respected participants would plead his case.

Work concerning theory was presented differently. A person (not necessarily a seminar participant) who wanted to present a paper on theory (either his own or a publication) had first to present it to Landau. If Landau agreed with the principal features of the presentation, it would be allowed for the seminar. In the course of the presentation Landau would give some clarifying comments and it happened quite often that his explanations strongly differed from the author's point of view. A loud discussion would follow. Quite often one could hear Landau exclaim: "The author himself does not understand what he has done!" In all cases Landau had an entirely original understanding of the work, and it was not easy for an ordinary person to follow his arguments. I was not the only one to need several hours, and sometimes several days, to realise the depth of his statements: they frequently showed a problem from an entirely new aspect.

If entrusted with a theory presentation, one no longer was obliged to give any compulsory reviews. For instance, Pomeranchuk never did any of these, he always reported on theory. Theory presentations were not made only by physicists of the Landau school but also by Tamm, Bogoliubov, Gelfand and many others. After the war and until 1955 no foreign physicists ever came to Moscow.

There were two exceptional members of the seminar community to whom general rules did not apply: Ginzburg and Migdal. Landau said once about Ginzburg: "Ginzburg is no pupil of mine, he just sneaked in". Ginzburg did indeed consider himself to be a pupil of Tamm. He was nonetheless one of the most active participants in Landau's seminars. He was not subject to the common rules of presenting overviews of articles, etc. He did however speak often and each of his lectures contained an abundance of new facts and new ideas, brilliantly and wittily presented. I still remember his lecture on supernovas which started with a historical introduction on observations of supernovas in ancient Babylon, Egypt and China. It is not by accident that the well-known phenomenological theory of superconductivity, the precursor of many modern models of spontaneous symmetry breaking, was created by Ginzburg and Landau.

The other exceptional person was Migdal. His name does not appear in the list of Landau's pupils — he did not pass the theorminimum but he was a full member of the seminar. Only Migdal was allowed to come late to seminars and nonetheless to enter the room through the front door. As a rule, seminars started punctually on time (to the minute) but sometimes Landau would say: "Let us wait another five minutes — these are Migdal's five minutes". Once in the middle of the seminar the front door of the hall opened and a figure clad in a fireman's uniform complete with fireman's helmet appeared. "Get out! Clear the room! We are starting fire drill exercises here!" said the man in a most decisive tone. Lifshits jumped up: "We hold a seminar here every Thursday! You have no right!". "Out! Clear the room!" said the man in an even more commanding voice. People started to get up and move to the doors. Then the fireman took off his helmet and also the thread which was pulling his nose upward. It was Migdal!

In 1958 Landau and a few members of the seminar were full of enthusiasm for Heisenberg's new theory, which stated that all elementary particles were produced from a universal fermion field. (Others, however, approached this theory with scepticism.) At one of the seminars Landau was handed a letter purporting to come from Pauli (obtained through the good offices of Pontecorvo) and Landau read it out loud. In this brief letter Pauli wrote that he liked Heisenberg's theory very much, he had found new arguments to support it and considered this theory to be very viable. Moreover, wrote Pauli, the latest experiments with Λ particles confirm Heisenberg's theory. The letter did not, however, give any data about these experiments. There was great excitement — Pauli was well known as a man of critical intelligence, he was definitely not gullible. A variety of hypotheses was put forward, one young theorist (L.B. Okun) even went to the blackboard and attempted to show what the experiment mentioned by Pauli might have looked like. Meanwhile, Migdal took the letter, read it attentively and said: "There is something odd here. If you read the first letter of each line from the bottom up, you get the Russian word "idiots". What might this mean?" The secret was simple: the letter had been written by Migdal and Pontecorvo.

Landau took part in the atom project, he worked conscientiously, at his own very high level. The reason for this was that he was afraid to refuse participation, although he did not like it at all. He already had some experience, having spent a year in prison (1938–1939) and he definitely did not wish to repeat the experiment.

This is what Landau did in the atom project:

1. He developed in detail the theory of neutron multiplication in the bomb, which made possible the calculations of its efficiency;
2. He developed the theory of a premature explosion of the bomb;
3. He participated in the work of Laboratory No. 3; in particular, Landau introduced the concept characterizing the uranium block in the reactor which is important for working out the theory of a heterogeneous reactor;
4. He worked out the theory of the shock wave caused by the bomb's explosion, taking account of radiation;
5. He took part (in the early stages) in work concerning RDS-6t ("the tube" — see below) and formulated the final conclusion concerning this system;
6. (Together with N.N. Meiman) he created the method of a numerical integration of a non-stationary dynamic problem, used for carrying out calculations of the processes happening in the bomb.

For his work on the atom project L.D. Landau was awarded the Order of Hero of Socialist Labour.

3.1.3. *A classification of theoretical physicists according to Landau*

Landau sorted all theorists into five classes, according to their achievements in science. Class 5 contained people whom he called "pathologists". This was of course Landau's style. He did not even consider people who published work that contradicted established laws of physics or well-known facts. He did however include into the relevant classes people who had made a mistake or made an incorrect assumption: he took into consideration the correct part of their work. The top rank in Landau's classification was of course held by Einstein. Landau put him into the class 1/2. (Classes with half-integral numbers were admitted.) The first class contained the creators of quantum mechanics: Bohr, Heisenberg, Schrödinger, Dirac, de Broglie. Many were surprised at his putting de Broglie into Class 1, but Landau retorted that de Broglie was the first to state that a particle could be simultaneously both a particle and a wave (i.e. under some circumstances it could demonstrate the properties of a point particle, and in others that of a wave). This idea is genius! The next class in Landau's assessment — class 1 1/2 — contained Pauli and Fermi.

Landau did not include Feynman in his lists of classes 1 or 1 1/2. This was due to the fact that the understanding of Feynman's work was greatly delayed in coming to Russia. For instance, at Landau's seminars two separate attempts were made to present the celebrated work of Feynman and in both cases Landau chased the lecturer off the podium. It was only the third attempt (by Pomeranchuk) that succeeded in breaking through.

Until 1937 Landau put himself into class 2 1/2. After creating in 1937 the theory of phase transitions he moved himself into class 2.

I consider that George Gamov ought to be allocated to class 2, and probably even higher. In my opinion he ought to have been awarded the Nobel prize, even two prizes — for physics and for biology. He was the first to calculate the probability of α decay as a consequence of the tunnel effect. The tunnel effect in quantum mechanics was already known, but this was its first application to a physical phenomenon.

Gamov was the first to propose that the Universe had started with a Big Bang. This ingenious idea was well ahead of its time. Now it lies at the foundation of all cosmological models. In genetics, after the discovery of the double helix, Gamov made the greatest discovery: 20 aminoacids link into the double helix. This too was an idea of genius. I advise everyone to read the George Gamov's book, *My World Line* [35]. (The book contains a large article by D.D. Ivanenko that takes up one third of the book. Ivanenko's statements about Gamov are probably correct, as to what he writes about himself ...)

I asked Ginzburg once to which class he himself belonged in his opinion. He said: "to the third". I said: "where do I belong then, to the fourth?" He immediately said: "you are right, I shall put myself into 2 1/2". In my considered opinion, the highest achievement of Ginzburg is the introduction to the theory of spontaneous symmetry breaking which is contained in the Landau-Ginzburg joint work on superconductivity (I know that this was done by Ginzburg, not by Landau).

I would put into class 2 1/2 also Gribov, Larkin, Keldysh. (Bronshtein ought to be in it as well, since he was the first to quantize gravitation, but alas, he was executed in 1937 at the age of 30.) I have named here only the people whose work I knew well. I do not discuss people of such renown as Tamm, Pomeranchuk and others, following the adage "bootmaker — do not judge higher than the boots".

Vladimir Naumovich Gribov was without any doubt the greatest theorist in Russia in the field of elementary particle physics. It is in fact he who created the Regge description of interactions at high energies. Landau had discovered his exceptional abilities well before anyone: in the 1950s, when Gribov's main work had not yet been done, Landau wrote in his comment about the possible award of the Lenin prize to Bogoliubov that the Lenin prize ought to be awarded to Gribov rather than to Bogoliubov. However, in the Nuclear Physics Division of the USSR Academy of Sciences, where Gribov was a Corresponding Member from 1972 up to 1997, eight scientists were elected to the full membership, but Gribov was not deemed worthy of this honour.

I consider that the greatest achievement of Anatoli Ivanovich Larkin is his discovery (in his collaboration with V.G. Vaks) of the spontaneous symmetry breaking in elementary particle physics, made in the 1950s. At that time it was not understood that spontaneous symmetry breaking in superconductivity and in particle physics are similar from the point of view of theory.

The reader may ask in which class the author puts himself? I shall answer: to the third. My highest achievements are the following: (1) Until the first half of the 1960s it was thought that strongly interacting particles, having a finite size, cannot in collisions absorb large momentum transfers. In my 1966 paper I showed that in some cases it is not so, i.e. they behave as if they were point particles. Using this result in an article written in collaboration with E.P. Shabalin, we showed that the theory of weak interactions, which contains only three quarks, has internal contradictions. Then Glashow, Ilipoulos and Mayann introduced a fourth quark into the theory and built a universal electroweak theory (S. Glashow: Nobel prize). (2) I demonstrated, based on the CPT theorem, that if parity is not conserved, then charge symmetry and/or time symmetry is also violated. (3) It was shown that in deep inelastic lepton-hadron scattering the interaction in space-time takes place near the light cone. (4) I established that the proton mass arises from spontaneous chiral symmetry violation in quantum chromodynamics, and calculated the proton mass with an accuracy of about 10%.

3.1.4. *History of some papers connected with Landau*

The Landau-Abrikosov-Khalatnikov paper. Immediately after joining ITEP (1950) I began to study renormalization theory by Feynman techniques. A.D. Galanin was attempting to calculate radiative corrections in quantum electrodynamics still using the old techniques. He changed to the new Feynman techniques and was in a way a senior colleague of mine. We learnt how to calculate radiative corrections in QED and meson theory, to carry out renormalization in the lowest order of perturbation theory, and then also for higher orders. I succeeded in constructing the exact system of coupled equations for the Green function in meson theory. Then, together with A.D. Galanin and I.Ya. Pomeranchuk, we renormalized the mass and charge of such a system. We could show that the solutions of this coupled system of equations must not contain infinities — they have to be finite. However, when cutting off this infinite system at some finite term, the infinities appeared again; in order to get rid of them one had to sum up the entire infinite series. Thus, this attempt did not succeed, but we did learn much.

Calculating together with Galanin the first few orders of perturbation theory, we saw that in the polarization operators of vertex functions at large virtualities p^2 terms of the form $\ln(p^2/m^2)$ arose: in the lowest order $\ln(p^2/m^2)$, in the second order terms proportional to $\ln^2(p^2/m^2)$, in the third order $\ln^3(p^2/m^2)$, and so on. We learned much from a paper by Edwards [28]. Edwards constructed an equation for the vertex function in tree approximation and found that in the nth order of perturbation theory terms $\left(e^2 \ln \left(p^2/m^2\right)\right)^n$ arise.

In the 50s, Landau used to come to TTL (ITEP)[9] every Wednesday. He took part — very actively — in the experimental Wednesday seminars organised by Alikhanov. After the seminar, Landau would come to the theorists' room which at that time housed Galanin, Rudik and me. All the other theorists would come there as well, and the discussions then would last for about two hours.

At one of these discussions, Pomeranchuk, Galanin and I explained to Landau the situation with radiative corrections in quantum electrodynamics. As a result of these conversations, Landau conceived the idea of summation of the leading logarithmic terms, i.e. terms $(e^2 \ln p^2)^n$ in QED. This is the reason why Pomeranchuk, Galanin and I were thanked in the first article by Landau, Abrikosov and Khalatnikov. (Landau was sparing with his thanks, he expressed them only to those people who had truly contributed something substantial to his work.)

At first, when Landau was formulating the idea, he thought that as a result of the summation of the leading logarithms in QED, the interaction decreases with increasing p^2 — nowadays known as

---

[9]TTL — Thermo-Technical Laboratory — was the old name of ITEP.

asymptotic freedom. Such an expectation was formulated in the first of a series of papers by Landau, Abrikosov and Khalatnikov, submitted for publication before the final result had been found. Visiting TTL on Wednesdays, Landau told us how the calculations were proceeding. The principal ideas — rotation of the path of integration, introduction of a cut-off, choice of a gauge — belonged to Landau. The technical calculations were done by Abrikosov and Khalatnikov — Landau himself had a poor command of the Feynman techniques. Their result confirmed their expectations — the effective charge in QED decreased with increasing energy.

Galanin and I decided to repeat these calculations. We wanted to include the same idea into our system of renormalized equations. (We later did this together with Pomeranchuk.) However, as early as in the calculation of the first loop, we came to the opposite result: the effective charge was not decreasing with the rise of energy, but rather it increased! On the following Wednesday we told Landau about this and persuaded him that we were correct. The last of the series of articles by Landau, Abrikosov and Khalatnikov, which its authors already meant to submit for publication, contained a mistake of a sign, which radically changed all conclusions — instead of the expected asymptotic freedom one got a null charge. In other words, quantum electrodynamics was shown to be an inconsistent theory. S.S. Gerstein (who at the time was working at the Institute for Physical Problems) told us later that Landau, on returning from this seminar at TTL, said: "Galanin and Ioffe saved me from disgrace".

A year or two after the publication of the Landau-Abrikosov-Khalatninkov articles and also of the Landau-Pomeranchuk article containing the more general justification of the null charge, Landau received a letter from Pauli. This said that a postgraduate student of Pauli, Walter Thirring, had found anexample of a theory which did not contain a null charge — a scalar meson-nucleon interaction theory. A preprint of Tirring's paper was included. Dau gave the paper to Chuk, and Chuk gave it to me, asking me to sort it out.[10] I studied the article and concluded that it was wrong. The mistake was

---

[10]Dau and Chuk are what the elder generation of Landau's and Pomeranchuk's pupils called them.

caused by the use of a Ward identity which arises in differentiating with respect to the nucleon mass, and was violated in the renormalization. I told Chuk about it. "You found the mistake, you must write to Pauli about it", said Chuk. I was afraid to write to the great Pauli saying that his postgrad had written an article with a mistake in it, and that he — Pauli — had not noticed it! But Chuk insisted and eventually I wrote to Pauli. The answer came not from Pauli but from Thirring. He fully acknowledged his mistake. The article was never published.

Work on the conservation of C, P, and T. In 1955–1956 everybody was worried about the θ–τ puzzle. Experimentally, decays of the K meson were seen into 2 and into 3 π mesons. With parity conservation, which at that time was considered to be unshakable, a meson could not decay both into 2 and into 3 π mesons. Therefore most physicists believed that they were two different mesons — θ and τ. With increasing experimental accuracy it became however clear that their masses coincided. In the spring of 1956 Lee and Yang published their revolutionary paper, in which they proposed the hypothesis of parity nonconservation in weak interactions, explained the θ–τ puzzle and calculated the effects of parity violation in β decay and in the chain of reactions $\pi \to \mu \to e$. Landau categorically rejected this possibility, saying: " Space cannot be asymmetrical!" Pomeranchuk preferred the hypothesis of parity-degenerate strange particle doublets.

A.P. Rudik and I decided to calculate yet another effect assuming parity violation, other than the ones considered by Lee and Yang. We chose the β–γ correlation. I made an estimate and found that the effect should be large. Rudik started detailed calculations. After some time he came to me and said: "You know, the effect is equal to zero". "Impossible!" I said. We sat down to check the calculations, and I noticed that Rudik, a well-bred theorist, when writing down the Lagrangian, had imposed C invariance (the particle-antiparticle symmetry). As a result, the constants multiplying the parity violating terms were purely imaginary. In Lee and Yang's work, these constants were arbitrary complex numbers. (If one puts them purely imaginary, then all parity violating effects disappear in their work too.) So, the question arose about the connection between the C- and P-invariance. I discussed this question with Volodia Sudakov, and in our conversation we remembered an article by Pauli. I had read it earlier but had entirely forgotten it. This was partly due to the fact that Landau was sceptical about this article: he thought that the CPT-theorem was just a trivial relation which any Lagrangian had to satisfy.[11] I should like to point out that in their article Lee and Yang did not say anything about the CPT-theorem or about a connection of C-, P- and T-invariance.

I reread Pauli's article, this time very carefully, and it immediately became clear that if P is violated, then the violation of either C or T, or both, is inevitable (according to the theorem proven by Pauli $CPT = 1$). And then the following thought arose: two K^0 mesons of very different lifetimes can exist only if either C or T (or both) are — at least approximately — invariant. Rudik and I considered several effects and saw that odd-P pair correlations of spin and momentum (terms $\sim \vec{\sigma}\vec{p}$) appear when C is violated and T is conserved; in the opposite case they are absent. (In a subsequent article I proved this theorem in its general form, and I also found the odd-P terms that correspond to T violation.) We wrote our article and showed it to L.B. Okun. Okun made a very useful remark that similar effects — different in schemes with C and T invariance — arise also in decays of K^0 mesons in π mesons. We included this remark into our article and I asked Okun to be our co-author. He refused at first, saying that such a remark deserved only an expression of thanks, but eventually I persuaded him [29]. After this we presented our work to Pomeranchuk. He decreed that this should be immediately presented to Dau on the very next Wednesday. To begin with Dau refused to listen at all: "I do not want to hear any talk about parity violation. It is nonsense!" Chuk pleaded with him: "Dau, be patient just for a quarter of an hour, do listen what the youngsters are saying". Dau eventually consented. I did not speak for long, about half an hour. Dau remained silent and left. On the following morning Pomeranchuk telephoned me: "Dau has solved the

---

[11] The CPT theorem is the statement that $CPT = 1$, where T is the time inversion.

problem of parity violation. Let us go and see him straight away".
By that time Landau had completed both his articles complete with
all calculations — one on the conservation of combined parity and
the other on the two-component neutrino [30]. Landau had assumed
that CP-conservation was an exact law of nature, which turned out
not to be true: Kronin *et al.*, in an experiment in the US on K^0
meson decay, established that CP is not conserved, but that CP
violation is small compared with the usual weak interaction [31].

Our paper and Landau's papers were submitted before the experiments of Wu *et al.* in which the asymmetry of electrons in the
β decay of polarized nuclei was shown — the correlation of the nuclear spin and the electron momentum was demonstrated, implying
the violation of parity. At the same time, our results indicated that
charge conjugation symmetry was not conserved in β decays either.
A corresponding note was added to our article. A similar statement
was also made in the article by Wu *et al.* in which the authors quoted the work of Lee, Oehme and Yang which had been done after us.
In their Nobel lectures, Lee and Yang referred to our priority in this
matter.

Unfortunately, the history of the work done by Landau on parity
conservation ended with a rather ugly episode which one does not
really want to describe, but one must since it did happen. Literally
a few days after Landau sent his articles to the *Journal of Experimental and Theoretical Physics*, he gave an interview to a Pravda
Corespondent, which was published straight away. In this interview
Landau talked about the problem of non-conservation of parity and
of his own way of resolving it. There was not one word about the
work of Lee and Yang (let alone of ours). All theorists at TTL were
indignant. Berestetsky and Ter-Martyrosian went to see Landau and
told him in detail what they thought about the interview. The result
of their action was the following: they were both expelled from the
seminar. I did not express my opinion to Landau in person, but I
did express it in conversations with those he worked with, and these
presumably told Landau about it. Landau punished me differently:
he removed my name from the list of those whom he thanked in
his article, he only left Okun and Rudik. This was something that

Pomeranchuk could not tolerate. He went to Landau and told him (I heard it from Chuk himself): "Boris is the one who explained everything about C, P and T to you. You would not have done this work without him, and now you are cutting him out of your acknowledgements!" I do not know what Landau answered, but I do know that he chose a compromise: he reinstated my name, but he did not put it in alphabetical order, I came second.

Landau considered CP conservation to be law of nature and did not admit its violation. In relation to CP he would say the same about spatial asymmetry as what he used to say about the violation of parity. I constructed an example of a Lagrangian in which T was violated while the vacuum was in no way affected, and I tried to make him change his mind, but he would not listen.

Applications of weak interaction theory. After 1958, when Marshak and Sudarshan and, independently, Gell-Mann and Feynman had formulated a universal four-fermion theory of weak interactions, I got interested in the question of higher order corrections to this theory which was not renormalizable. The idea was the following: as a result of higher order corrections to weak interaction there should be a number of observable effects. Their absence in experiments would show the upper limit of validity of the theory of weak interactions. It was assumed that the integration over the virtual hadron momenta is cut off by the strong interactions, and therefore diagrams with virtual hadrons need not be considered — their contribution is small. In the 1960 paper [32] two such effects were considered: the decays $\mu \rightarrow e + \gamma$ and $\mu \rightarrow 3e$, and corrections which violated the equality of the μ and β decay constants. At that time it was assumed that there was only one neutrino, and that therefore the decay $\mu \rightarrow e + \gamma$ was allowed. The strongest limit came from that decay: $\Lambda \geq 50\,\mathrm{GeV}$. But when it was established that electron neutrino and muon neutrino were different, this limit was discarded. Since the study of purely leptonic processes did not produce any limits, and since there was consensus that processes with virtual hadrons were cut off by strong interactions, it was felt that this path did not open any perspectives.

In my 1966 article [33] I established that due to current algebra there were some cases[12] where the strong interaction did not cut off the virtual hadron amplitude. E.P. Shabalin asked me the question: might it be possible to use this technique in studying weak neutral currents, in which experimental limits were very strong. Together with Shabalin [34] we considered the decay $K_L \to \mu^+\mu^-$ and the mass difference of K_L and K_S at order $G^2\Lambda^2$ in a theory with only usual and strange quarks (i.e. only u, d, and s quarks). We showed that by virtue of current algebra, there is no cut-off of virtual weak interactions by strong interactions, and we calculated the amplitude of $K_L \to \mu^+\mu^-$ and the mass difference $K_L - K_S$ to order $G^2\Lambda^2$. The strongest limit on the cut-off, $\Lambda \leq 5\,\text{GeV}$ came from the mass difference $K_L - K_S$. In the quark language that meant that the weak interaction theory with u, d, and s quarks changes its form at fairly low energies $E \leq 5\,\text{GeV}$. This statement was the starting point for the hypothesis of Glashow, Iliopoulos and Maiani of the existence of the c quark and the formulation of the weak interaction theory in a form where the contribution of the c quark compensates the divergent terms due to the u, d, and s quarks (GIM mechanism).

We reported on this work at an ITEP seminar, issued a preprint, the article was published in the journal *Yadernaya Fizika*. A little later, L.B. Okun attended a seminar in the US. On his return he reported at an ITEP theoretical seminar on everything new he had learnt at the conference. The principal news was a relation for the mass difference of the K_L and K_S mesons found by Gell-Mann, Goldberger, Low and Kroll from weak interaction corrections. The formula he wrote on the blackboard, I recognized as our formula — they coincided exactly. (Strictly speaking, as in our formula, so also in theirs, saturation by the vacuum state was assumed. We had, in addition, calculated the contribution of the one-pion state and shown that it was small.) After the seminar I showed Lew Borisovich our published paper and reminded him that he had been present at the seminar at which I had reported about it. He suggested that I should send a letter and a reprint of our paper to Low, who had reported

---

[12]The paper was about corrections to the β decay constant in weak interactions in a theory with an intermediate vector boson.

on their work at the conference. In his reply, Low agreed that we had done the same work as they had done, and much earlier. The paper of Gell-Mann, Goldberger, Low and Kroll never appeared in print. In a review paper on that subject, Low referred to our paper and did not mention their work.

In connection with this activity, I received an invitation from Marshak to give a talk at the 1967 International conference "Particles and Fields" in Rochester. There is an interesting story to the gestation of this invitation. In summer 1967 Marshak took part in a conference organised in Yalta by the Ukrainian Academy of Sciences. Its Organising Committee was chaired by N.N. Bogoliubov, the conference was run at the highest level from the point of view of comfort, service etc. A strict hierarchy was observed among those invited: some were allocated the best black caviar, some received red caviar, some were just issued nice sausage.

Gribov and I were also taking part in this conference and we had many discussions with Marshak. (I had met Marshak as early as 1956 when he first visited Moscow.) During one of these discussions Marshak said that he would send us both an invitation to present talks at a "Particles and Fields" conference he was organising. "But they won't let us go to it!" said Gribov and I. "I do understand all that" — said Marshak — "but I shall trick them into it: I shall also invite Bogoliubov and I shall make the invitation conditional — all three must come. If they won't let you come, I shall recall Bogoliubov's invitation".

Marshak's trick worked. I started going through the clearing procedures and reached a very high level, one I had never reached before. I composed a paper for the conference in which I wrote up our results on the $K_L \to \mu^+\mu^-$ decay, the $K_L - K_S$ mass difference and several others. At the last minute I was not allowed to go to the conference. But they did allow Gribov to go. Marshak found himself in a difficult situation — he could recall Bogoliubov's invitation, but that meant Gribov could not come either. He cut the pear in half so that Gribov went to the conference but I did not. Gribov agreed to present my paper, but he did not like the piece about the $K_L - K_S$ mass difference, and in particular the assumption of the vacuum saturation, and demanded that I should cut out that piece. I had to consent. As a result this part was less known in the West, and that led to Marshak, Mohapatra and Rao doing similar work, though much later.

3.1.5. *The personality of Landau*

Many have written about Landau (for instance, see [10, 11, 24–27]) I shall try to avoid repetition and will only try to talk here about what in my opinion has not been discussed sufficiently. The thing is that in creating his school, in creating his seminar and in much else Landau had only one aim — to maintain Physics at a high scientific level. He did not attach importance to his having his own school, to having a large number of pupils who venerated him as "maitre" (Pomeranchuk sometimes called him that). What he wanted was to have his pupils always to remain at the cutting edge of science. He did not at all want — indeed, it was alien to his nature — his pupils to forge a career in science, to occupy leading, directorial posts. After the terrible car accident, when Landau barely reacted to his surroundings, somebody came to him once and said: "Dau, your pupil is now a Director". "A pupil of mine" — said Dau — "cannot be a Director!" External signs of servility towards The Teacher were alien to Landau, so much so that I cannot even imagine what he would have done if someone had shown these qualities. I expect he would have thrown them out. In this he was the direct opposite to some leaders of other schools.

Landau felt his own responsibility — something like "the white man's burden" for the level of science to be kept high. He did not keep silent — as most people do now and as is usual in the West — when a person in a presentation made wrong statements in his presence. The very existence of Landau kept this level high — hardly anyone dared present a raw idea, an idea that had not been thought through, because people were afraid of Landau's critique. Pomeranchuk told me once "You cannot imagine what an immense sanitising job Dau has done in theoretical physics". If for some reason Dau did not want to criticise an author in public, he just did not come to his lecture. This is what happened with Rumer's paper on five-optics, which Rumer gave once he was back in Moscow after many years in prison and exile. Dau was fond of Rumer but he did not consider five-optics correct and therefore he did not come to the lecture. E.L. Feinberg gives an excellent description of this episode [10].

Landau clearly identified problems which he was capable of solving and those which he could not solve. A typical saying of Landau's is: "How can you solve a problem if you do not know its answer in advance?" He restricted himself to a definite class of problems and in this class Landau experienced no difficulty in solving problems, the difficulties arose only in identifying and setting them. Landau would not consider problems for which he had no answers in advance, and this was not only his strength, but also his weakness. He was thereby refusing to attempt the solving of problems which in his opinion were above his class. It seems to me that as a result of such modesty in self-assessment Landau did not achieve all that he could have achieved (in particular concerning quantum field theory).

In his whole behaviour Landau did not conform at all to the generally accepted image of a dignified academician. Pomeranchuk could say to him: "Dau, you are talking nonsense!" and Landau would take this quite calmly, though of course he demanded that this statement be proved correct.

Right from the founding of ITEP, Landau would regularly come every Wednesday to the experimental seminar run by Alikhanov, and afterit he would stay behind for an hour or an hour and a half for talks with theorists. Right through 1946 he was the head of the Theory laboratory, after this Pomeranchuk became the head and Landau remained a part-time employee of the lab. This went on until 1958, when holding multiple jobs was prohibited, and Landau was dismissed. (Zel'dovich, who was also a part-timer, was dismissed at the same time as Landau.) These Wednesday conversations included all theorists (we were very few at the time, only 5 to 7 people). They were very diverse and free, included all sorts of subjects — from physics to politics and literature, and for us (the young ones) they were fascinating.

Now back to Landau's personality: I think that an internal

modesty, or rather even a diffidence, was presumably a trait of his character. He understood his weakness, he tried to fight it — particularly in his youth — but this would turn into showing off. I do not agree with E.L. Feinberg [10] that he sort of had two faces (I do not like the word "mask" used in Feinberg's book): on the outside he was cutting and provocative, while on the inside he was soft, diffident, easily hurt. This duality affected Landau's relationships with women, which are described (but in a very distorted way!) by his wife Cora Drobintseva [24]. (In my opinion, this book is disgusting. To show the character of the author I shall give just one fact: after the traffic accident, when Landau was hanging between life and death, physicists organised a team whose members were present round the clock in his hospital, who organised the provision of medication, doctors, special foods, etc. They received instructions from E.M. Lifshits and his wife Elena — these two in fact were in command of the struggle to save Landau's life. The team worked in shifts, each shift had a leader who was responsible for his team. I was one of these and was on shift in hospital every 3–4 days. I can state categorically that Cora did not appear even once in the hospital over the first month and a half. Not once! She appeared for the first time six weeks after the accident, when it became clear that Landau would live.)

In her book Cora presents Landau as a sort of Don Juan, and maybe worse. It seems to me that although she lived with Landau for many years, she never worked out what her husband was like. This has been done much better by A.S. Kronrod. He once presented Landau to a lady who may not have been a woman of loose behaviour, but who was definitely very close to it. After a certain time Kronrod enquired: "Well, did you manage to make it with this lady?", "Oh, no," — said Dau —" You know how touchy and reserved she is!" Kronrod explained this story thus: "It was not the lady who was touchy and reserved, it was Landau who was diffident inside, and the lady — being experienced — felt this immediately".

Now about something else. In conversations and at seminars Landau liked to talk in aphorisms, these were his own creations and he would sometimes repeat them. Many people treated his aphorisms with contempt: "There he goes again, playing the same old record!"

In reality, however, there was deep sense in these aphorisms. Landau realised that no one disputes an aphoristic statement, and its use was the best method for closing a pointless discussion, which otherwise would last forever. I shall give a few examples. Landau was aware that much in life (and in science) was stupid, and only little was reasonable. The relevant aphorism was: "Why are singers stupid? Because their selection is ruled by other factors". And here is another one which, by the way, is very relevant for our current times: "When hearing about some extraordinary phenomenon, people begin to try and explain it by proffering implausible hypotheses. Before all else, do consider the easier explanation: that this is all sheer lies".

And lastly — and this is extremely valid for our times — Landau was sure that it is compulsory for a leader in science to have his own results in research, the value of which is universally recognised. Only then does he have the moral right to manage people and to set them problems to solve. (And I will say for myself that this also concerns his right to give advice to politicians.) Landau would say: "One cannot build a career in science exclusively on one's probity — this leads inevitably to losing both science and probity". One does feel like generalising: one cannot build a career in science exclusively on one's management skills: the consequences would be similar.

3.1.6. *The leaflet*

On 27 April 1938 Landau was arrested. He spent exactly one year in the Lubyanka prison of the NKVD — he was freed on 28 April 1939 under guarantee from P.L. Kapitsa. Landau suffered disastrous weight loss in prison: he could not eat prison food. As he would say: "If I had spent another 2 months in prison, I would be dead". (I heard these words from him myself.) This imprisonment affected his character and behaviour for the rest of his life. The historian G.E. Gorelik succeeded in gaining access to Landau's file in the files of the KGB [25] and found that the main charge against him was "participation in an anti-Soviet group which existed in the Kharkov Institute of Physics and Technology, and in sabotage". But the file also contains a leaflet written by M.A. Korets, a friend of Landau's.

Landau had approved this leaflet. He confessed this in statements made in prison. The leaflet, written in late April 1938, purported to come from the Anti-Fascist Labour Party which in fact did not exist. The leaflet was meant to be distributed during the May Day celebrations. The content of the leaflet is staggering: It says that "Stalin's clique carried out a fascist coup d'état", Stalin is compared to Hitler and Mussolini, the leaflet appeals to workers to overthrow the Fascist dictator and his clique, to join the (non-existent) Anti-Fascist Labour Party. (The full text of the leaflet is given in Gorelik's publication [27], and an abbreviated version in Feinberg's book [10].)

Landau's file contains only a typewritten copy of the leaflet, Gorelik did not see an original. The times in 1938 were such that a leaflet of this sort would cause the immediate execution of people connected with it (people were shot for less), but in this particular case Landau was freed after one year and even Korets received a relatively mild sentence (10 years of labour camp, which was extended by another 10 years) and after this he lived in freedom until his death in the 1980s. In Landau's file this leaflet is not part of the main charge (sabotage) but only a supplement to it. One may well ask — might not the leaflet be a fabrication by the NKVD? Of course no-one doubts the mastery of NKVD in such matters. Gorelik [25] gives a negative answer to this question, Feinberg agrees with him [10], for Ginzburg [11] the matter remains unclear.

The main argument of Gorelik is that in the 1930s Landau was a communist in his life philosophy: he was certain that science could develop successfully only under a communist regime, only when the path to science was open for all talents from all layers of society. According to Gorelik, seeing the mass arrests in 1937, Landau understood that the existing regime had nothing in common with the one that he himself envisaged, and he decided to fight against it. This thought looks naïve to me. Landau was first and foremost a man of rational thought. He introduced a rational approach to everything, even into places where it ought not to be introduced — the whole story of his life shows this. He knew perfectly well that his closest friends — M.P. Bronstein, L.V. Shubnikov — and many other physicists had been arrested in 1937 and disappeared without trace in the cellars of the NKVD.

Landau moved from Kharkov to Moscow in February 1937 precisely to escape his inevitable arrest (he suspected that he was under surveillance in Moscow as well — see [10]). He could not fail to know that the leaflet, if distributed (in fact it only existed — if it existed at all — in one copy) could only lead to the arrest of its authors. As to their ensuing fate — well, there was no hope at all for them. Only a man who wanted to become a martyr would undertake the creation and distribution of such a leaflet. Landau did definitely not belong to this type of people — he was, if anything, a hedonist. And in any case Landau was of course not stupid enough not to foresee all consequences. Therefore I do not think that Landau could to any extent take part in composing such a leaflet. I also do not think that Korets would have come to him with such a proposition: according to the testimony of people who knew Korets, he liked Landau and, of course, must have been aware of the danger to which he was subjecting him.

I feel that the following possibility is more likely: in prison, Korets was forced to sign a confession saying that he wrote the leaflet and that Landau approved its text. It is also likely that in reality the text of the leaflet was written by the investigator. Subsequently Landau (who had been broken down by that time) was shown the leaflet and Korets' confession and he also confessed that he had taken part in composing this leaflet. All this does not look surprising, since Landau wrote in his own hand statements in which he admits his anti-Soviet activity and much, much more. (The NKVD were masters in beating out confessions. In a well-known case, one of their prisoners confessed that he had blown up a bridge across the Volga river. Many years later, after returning to freedom, he saw the place in question and verified that that particular bridge was intact — no one had ever blown it up.) In the 1970s, when he was already free, Korets began to speak out: he said that he had indeed written the leaflet. It is however possible that it was easier for him to say this than to admit that he wrote his confession under torture. When Korets was released in the 1950s, Landau did give him material support. Here too there is no contradiction — Landau knew from his own experience by what means Korets had been forced to make his confession.

The question remains: why did the NKVD need this leaflet which was not used as part of the main charge? A variety of hypotheses is possible. Here is one of them. At the start, several approaches were envisaged, one of them being a big trial, possibly even a show trial of scientists. This needed some weighty proofs of guilt. Later this thought was discarded. Maybe Stalin was in some way influenced by the letters in defence of Landau sent to him by Western scientists. Maybe politics underwent a change, with a small thaw at the time when Yezhov was replaced by Beria: Landau was arrested under Yezhov, his statements are dated 8 August 1938 (still under Yezhov) and his case was presumably meant to be dealt with at a later date. (For comparison: Beria was appointed Deputy People's Commissar of the NKVD in August 1938 and Commissar in November of the same year.) What is the truth? The secrets of Lubyanka remain secrets to this day.

3.2. I.Ya. Pomeranchuk

3.2.1. *Pomeranchuk's principles*

I first met I.Ya. Pomeranchuk in the winter of 1947/48. I was in my fourth year at the University's Department of Structure of Matter. It was time to look for a supervisor for my diploma research. I wanted to choose him from the school of Landau (this was allowed at the Department of Structure of Matter but not in other departments of the Faculty of Physics). I chose Pomeranchuk for the simple reason that by that time all other potential supervisors (V.L. Ginzburg, A.S. Kompaneyets) had already been booked. I did not know Pomeranchuk before, I had never seen him and I did not even know on what he was working. I rang him up, introduced myself and said that I had already been successful at three parts of the Landau minimum. This proved to be sufficient and he invited me to his house for further discussion. I was struck by the fact that there was practically no furniture in his flat: a desk in one room and in the other a camp bed covered with a grey army blanket, on top of which lay F.

I.Ya. Pomeranchuk

Bloch's "Theory of Magnetism". Every 3–5 minutes Pomeranchuk looked at his watch, bringing it close to his eyes — he must have been short-sighted. I asked him: "Am I keeping you from something?" — "Don't mind me," — he answered — "it is just a habit". Our conversation was brief. It ended with Pomeranchuk saying that he was accepting me for my diploma on condition that I complete the Landau examinations.

In winter 1948/49 Pomeranchuk was giving his "Quantum Field Theory" lecture course at the Moscow Institute of Physics and Technology (MIPT). This included: quantization of the electromagnetic field, Fock's method of functionals, theory of radiation, the Bloch-Nordsieck method, multi-time formalism, etc. This course was unique in all respects: one could neither hear it anywhere else nor read it anywhere else in such a complete form. I had the luck of attending part of this course and although I was far from understanding everything I still remember a feeling of clarity and of delight at the beauty of theory.

In 1949 Pomeranchuk was also giving a course at Moscow State University (MSU): neutron physics and theory of nuclear reactors. This amalgamation of high and low "styles", a mixture of abstract theory and concrete, even applied, physics was always a feature of his work and he was trying to graft this approach onto his pupils. Thus, while still working on my diploma, I was given two completely different problems: one was called "Production of polarized neutrons and their depolarization at moderation", and the other "Dependence of cross sections of bremsstrahlung and of pair annihilation on the photon polarization".

From 1st January 1950, I started work at the Laboratory of Theoretical Physics of ITEP. Very soon I got to know the principles which Pomeranchuk had set as a foundation to the work at this laboratory. Here are these principles:

1. "The management must be respected". This meant that all tasks set by the management for solving applied problems must be carried out first, with full conscientiousness and with a guarantee of faultlessness.

2. "Experimentalists must be respected". This meant that if an
   experimentalist comes to the theory group with a question or
   with a request for help, his question must be answered, his
   request must be satisfied, and, if necessary, calculations —
   even complex ones — must be carried out.
3. "No preferential treatment here". Everything is equally important. This means that every member of the laboratory
   must carry out the tasks set by the management or by experimentalists in equal measure. (Alas, as frequently happens
   with good principles, this one was not followed as strictly as
   the previous two.)
4. "Scientific work can be carried out between 8 pm and mid-
   night". This meant that the staff (especially the young
   members), in spite of being fully occupied by work set by
   management and experimentalists, must find time to engage
   in high science.
5. "The Institute is not a charitable organisation". In other
   words, the Laboratory (and this extended to the whole Institute) may not tolerate staff who do not work well or well
   enough.

This latter demand of Pomeranchuk coincided exactly with Landau's attitude. I once heard Landau talk to Pomeranchuk about
Sudakov (Sudakov was possibly the most talented theorist of the
post-war generation at ITEP): "Sudakov is work-shy. These tall
chubby blond men are often work-shy. You carry on pushing him,
make him work, don't allow him to shy away from work!"

Pomeranchuk did not reduce his demands on his Laboratory's
staff even when their status rose — when they reached the status of
Doctor of Science. On the contrary, these demands rose. Here is an
example. From the mid-fifties Pomeranchuk had me write the annual
report on the Laboratory's work (on non-secret matters). At that
time the staff was relatively small, I was familiar with most of the
work that had been done, and if I was not, I would ask authors to
hand me their information on paper (whether published or not) and
I would study these. For the report I used my own judgement on the

value of these articles: some received more attention and others less, and some did not appear in the text of the report at all, they were only mentioned in the list of work undertaken that year. Once I had completed this job, I would take the manuscript to Pomeranchuk. He would read the text attentively in my presence, cross something out and add something else. Sometimes he would say: "This needs correcting — add ...". After that I would hand the corrected text to the typist. Once the report was typed, Pomeranchuk would sign it unread. My work and my assessment obviously satisfied him on the whole, since this practice continued till the end of his life.

In 1963 (maybe 1964, I cannot remember exactly) I came to Pomeranchuk as usual and brought the annual report. This contained among others a description of research carried out by two members of staff of our Laboratory who both had DSc degrees. I had put them into the second category, in which results are described, but briefly. Pomeranchuk read the report and upon reaching the passage on this research crossed out its description with a fat cross and said in a harsh tone: "I do not understand what these people are doing! This is not science! As Director of the Laboratory, I cannot be responsible for what they are doing. I shall raise this with the Director of the Institute: they must no longer remain in my Laboratory". And within a short space of time two new theory laboratories were created, headed by these two Doctors of Science.

Pomeranchuk was however not a slave to his principles and broke them when necessary. In 1952 Rudik and I had to sit the Philosophy part of the PhD examination. In those days it was a most serious task: one had to know by heart a huge number of quotations, and the least departure from the text resulted in a drop in one's assessment. To get a mere "pass" in it was equivalent to disaster. Pomeranchuk decreed that we were to stop all work for two weeks, we were not to appear in our room so that no-one could find us, we were to sit in another part of the university building and study philosophy. After this we successfully passed the examination.

Pomaranchuk and I turned out to have a common interest outside science — we read newspapers. This was an interest that the rest of our staff did not share, and in any case hardly anyone read any papers at all in those days, since they did not contain any information. The newspapers were all full of articles which started with words like "A new level of achievement was attained in the production of ..." and "The milling-machine operator Ivanov (rolling-mill operator Petrov, etc.), using Stakhanov's methods, achieved in one shift a level of 10 (20, 50, ...) times the norm". If one wanted to extract any information from a newspaper, one had to be a true specialist in the matter, and we — Chuk and I — were true specialists. (From now on I shall call him Chuk, as many people did.) Each morning, as soon as Chuk arrived at ITEP, he would come to my room and ask:

> – Have you read today's "Pravda"?
> – Yes, I have (I would answer).
> – And did anything attract your attention?
> – A brief note on page 3.
> – Oh! — Chuk would lift his index finger.
> – What about you?
> – I suppose the same as you, a bit about the Voronezh Region plenary?
> – Yes.
> – And what attracted your interest in this report?
> – The greetings to members of the Politburo.
> – Oh! — Up went the finger again.
> – What exactly?
> – The order in which the Politburo were listed.

We understood each other well. From this order one could deduce which Politburo member was on the up and which was going down, and consequently one could deduce what would be happening to us.

But one morning, while still at home, I read in the newspaper that Pomeranchuk, together with Ivanenko and Sokolov, had been awarded the Stalin prize for their work on synchrotron radiation. On arrival at work I went to my room (at that time — it was at the very start of my work in ITEP — I was located in a different building, not in the one where all theorists were: Pomeranchuk had "lent me out" for a while to V.V. Vladimirsky to carry out calculations of the field in a linear accelerator). Later on, around midday, I went to the theorists' room which was usually occupied by Berestetsky, Galanin and Rudik, and this is the picture that I saw: three men were sitting at three desks, each writing something in his own documents. Pomeranchuk, Galanin and Rudik. I found myself in a difficult situation: I knew that Chuk had done some research with Ivanenko on betatron synchrotron radiation for which he had been given the Stalin prize. But I also knew that Landau could not stand Ivanenko, he maintained that Ivanenko had no theoretical achievements at all[13] and, moreover, when Chuk had collaborated with Ivanenko, Landau had expelled Chuk from the theory seminars and this disgrace lasted a considerable time before Chuk was eventually forgiven. I could not work out what had happened before my arrival, or whether the others had congratulated Chuk, or what I was meant to do. (I later found out the following from Rudik: he was in the room with Galanin. The door opened, Chuk entered, went quickly to Berestetsky's desk, briefly greeted the others, and without saying a word, sat down and began to write something. They did the same and this continued until I came.) I went up to Chuk and said: "Isaac Yakovlevich, I do not know whether one has to congratulate you or to express one's sympathy". Upon these words, Chuk unfroze and told us the history of this unfortunate collaboration.

"At that time (1944), said Chuk, I usually had lunch at the House of Scientists. One day I happened to be at the same table with Ivanenko. In those days we had a perfectly normal relationship and I described to him my research on synchrotron radiation which I was just finishing at the time. Ivanenko listened but did not make any comments. Some time later I met him again in the House of Scientists dining room and he said to me: 'Last time we discussed the problem of synchrotron radiation. We ought to publish an article about it, you and I'. I did not have the guts to say: 'But what have YOU to do with it?' That's how this article appeared." (Ivanenko describes

---

[13]It is usually said that the main achievement of Ivanenko is his paper on nuclei consisting of protons and neutrons. (Before neutrons were discovered one had to think that because the atomic weight of nuclei does not correspond to their charge, the nuclei contain electrons, and that produced many contradictions.) Landau used to say about this publication: "At that time (once Chadwick had discovered the neutron) everybody understood that nuclei do contain neutrons, and only Ivanenko went and published this."

the origin and development of this article quite differently [35, 36].)

The 1940s, particularly their second half, were very fruitful for Pomeranchuk: they include the foundation of the nuclear reactor theory, the theory of liquid ^{3}He and much more. But field theory always came first for him. He considered all those "Pomeranchuk effects" which he thought up with such brilliance to be his weakness and he frequently criticized himself bitterly for them.

In 1950, all the staff at the ITEP Laboratory of Theoretical Physics — Berestetsky, Galanin, Rudik and I — were intensely studying new methods in quantum electrodynamics: articles by Feynman, Schwinger, Dyson and others. Galanin and I translated some of these papers into Russian and they were published in review collections.[14] Pomeranchuk was greatly in favour of this activity but he hardly took part in it before the middle of 1951: in 1950–51 he was sent to Arzamas for 6 months to work on the hydrogen bomb project.

In 1949–1950 few people in Moscow understood that the creation of renormalization had opened a new era in particle physics and quantum field theory. (Apart from our group there were Abrikosov, Khalatnikov, Fradkin and maybe one or two more people.) The majority thought that because infinities remained in the theory, the renormalization theory was only just an attempt to push the dust under the carpet. Pomeranchuk did not share this opinion. He maintained that although renormalized quantum electrodynamics (or meson theory) was not yet a new theory, it was nonetheless a very important step towards it. And he awaited this new theory with impatience. "When the new theory arrives — he used to say — we will move into army barracks and we will put on army boots. Dr Ioffe, — he would ask sternly, — have you got boots?" I had to admit that I had not, I only had army half-boots with gaiters. Chuk was willing to accept even this, just so that the new theory would arrive. Landau was sceptical towards new developments in quantum electrodynamics: he did not believe that the difficulties with infinities could be circumvented by renormalizing mass and charge. At Landau's seminars, "alum"

---

[14]In 1949–51 we had difficulties in accessing American journals on physics, they often came very late and sometimes they bore a stamp "classified!"; we knew that they came to us illegally, via Sweden.

continued to reign supreme.

Landau considered me to be snobbish. He repeated many times: "Boris is a snob". By this he meant that I did not want to engage in solving real problems of physics but chose refined theory for preference. His words did not affect Pomeranchuk, to whom they were most frequently addressed, since Chuk and I were as one in this. But — and this was worst of all — Landau would also say this to Alikhanov, the Director of ITEP. And in Alikhanov's eyes Landau was an absolute authority in matters concerning theory and the assessment of theorists. Therefore Landau's comments could result in consequences which would be most unfavourable to me. Fortunately, Alikhanov had his own opinion on this subject: he knew perfectly well that I was doing the calculations for nuclear reactors and for the experimental apparatus needed for his — Alikhanov's — own research, so he knew that I was definitely not a snob.

Pomeranchuk got involved in the study of renormalization theory in 1951 and brought his typical passion to this. One morning Chuk rushed into the room where Rudik and I were working. He was in a state of absolute fury. I never saw him as furious as that — neither before nor later. He was shouting: "You read Feynman, you read Dyson, and you have not understood a single thing!" It took some time before we understood what the matter was. Pomeranchuk had understood that when calculating Feynman integrals one had to take the residues of the propagator poles such that some of the p^2 were equal to m^2. This led him to conclude that Dyson's method of finding the degree of divergence of diagrams by counting the powers of momenta was wrong, and we had not noticed that. Only in the evening, when Pomeranchuk had cooled down a bit, did we succeed in persuading him that Dyson was right after all. After this Pomeranchuk never allowed himself such explosions against us.

3.2.2. *Pomeranchuk's seminar*

Pomeranchuk tried time and again to persuade Landau to shift the area of his interests towards quantum electrodynamics and meson theories. He repeated again and again: "Dau, there are masses of problems here. They are difficult, they are just right for a man of your class". But Landau retorted each time: "I know my capabilities, the solution of infinities is not for me!"

In autumn 1951, Pomarachuk organised a seminar on quantum field theory and elementary particle physics. The seminar could not be held at ITEP, because not all participants had a pass for the ITEP site. Therefore Pomeranchuk arranged with Landau that his seminar would be held in the conference hall of the Institute of Physical Problems on the same day as Landau's seminar — on Thursdays — but two hours earlier. Pomeranchuk appointed me to be the seminar secretary. The first session was held on 1 October 1951. I reported about Dyson's publication at that seminar. Alikhanov, the Director of ITEP, asked me to write up an official report about the organisation of the seminar, and this I did. This document has been preserved. Practically all theorists of repute and very many young scientists took part.

At his seminars, Pomeranchuk demanded from the speakers first of all a clear and precise presentation of the physics: the speaker needed to have a full understanding of the physical meaning of the problem and of the results obtained. One had to consider limiting cases and compare them with results obtained by other methods (if such existed). Once A.M. Baldin was reporting at a Pomeranchuk seminar on his calculations of cross sections of pion photoproduction on nucleons in perturbation theory. He was writing long formulæ on the board and showed many graphs. Pomeranchuk asked him: "What is the behavior of the cross section near threshold?" Baldin showed one of his graphs. "And what is the asymptotic high-energy behaviour?" — Baldin showed another graph. "The showing of graphs is not a discussion method in theoretical physics" — shouted Pomeranchuk — "your presentation is over!"

In 1953, when the security restrictions of entry into ITEP were becoming less strict, the seminar was transferred to ITEP. In the form started by Pomeranchuk, the seminar continued until 2005.

3.2.3. *Pomeranchuk outside science*

Pomeranchuk barely existed outside science but one cannot say that he was completely non-existent. He said of himself (sf. "Memoirs" by A.D. Sakharov [5]): "I am an old-fashioned man, and for me some strange things still are the most important — things like love, for instance". And I will add friendship to this. There were several people whose names always evoked warmth in Chuk's voice — Shmushkevich, Akhiezer, Migdal. Chuk would exclaim "Ilya Mironovich!" (Shmushkevich), his index finger would rise, and his intonation implied that Ilya Mironovich was obviously deemed capable of such deeds (mostly of a debauched nature) of which no one else was capable. If Ilya Mironovich (an imposing man of great presence) was present on such occasions, he would turn brick-red. (In fact, Ilya Mironovich was a highly virtuous man, completely incapable of any debauchery whatsoever.)

Chuk's attitude towards Landau was different: one cannot talk about God with warm affection, and for Chuk Landau was indeed God.

They say that in his early years Chuk was a right-thinking Soviet citizen and member of the Komsomol. Not an active one, of course — all his active energy went into science — but right-thinking all the same. When Landau was arrested, this was an unbearable shock for Chuk (God had been arrested!). From that moment Chuk became very prudent (excessively prudent on many occasions). In 1956, after the forcible disbanding of ITEP's Party organisation, there was a campaign at the Institute, designed to consolidate the newly reformed Party organisation: it needed new members and these were actively sought. One day Chuk came to me and said: "Boris Lazarevich, I am not trying to talk you into joining the Party, but if you did this, I would accept it with understanding". Chuk knew my political opinions very well indeed and these were very different from communist ones.

On that occasion he also rang Nikitin and said:

– Sergei Yakovlevich, may I please come and see you?

– Dear Isaak Yakovlevich, please do not bother, I shall come to you myself.

This is what Nikitin tells about the meeting:

When he arrived at Chuk's office, he found Gribov already there.
– Volodia, said Chuk, may I ask you to leave us for a minute?

When we were alone, Chuk said:
– Sergei Yakovlevich, in my opinion you ought to become a Member of the Party.
– Only after you, Isaak Yakovlevich, I said.

Chuk remained silent for a little while.
– Right, he said, as I understand you, we can call Volodia back?

Chuk had an assortment of his most favourite jokes which fitted many life situations. Here is one of them.

There was a man who lived and worked on the outskirts of Moscow but who frequently needed to come to the town centre. And there he would use a certain public toilet. Since he did this often, he became friendly with a middle-aged woman who worked there. Sometimes he would give her some small present. Then he had to leave Moscow for a certain time. On his return he visited the public toilet again, but his friend was no longer there. He was sad but thought: She was after all an elderly woman, anything might have happened...". After some time he happened to visit a public toilet on the outskirts of the capital — and what joy — there he saw his friend. "Why are you here and not in your former place?", he asked — "There were intrigues", she replied.

This story is eternally valid, and I think that right now it is even more valid than in Chuk's time.

And here is another of his favourites: There was a village with a priest and a mayor who detested each other. Once the priest was walking along the river and saw the mayor fishing. The priest thought: "Right, I shall come up to him and ask 'Well, how is the fishing going?', — If he answers 'It is good!', I shall say 'Yes, any fool can catch fish from this site!', and if he says 'It is bad', I shall say 'Even the greatest fool would not try to catch fish here'". So the priest asks: "Well, how is the fishing going?" — "Now, you go and ..." answers the mayor.

And Chuk would conclude: in any situation there is always a third alternative.

Chuk remembered one day when his toilet at home broke down. He called a plumber. An old man turned up and started mending the toilet. Half an hour went by, then an hour, and still nothing happened. Chuk came up and said: "What about trying this here and that there?" — "Now don't you start giving me advice," — answered the plumber — "I have been working in shit business for thirty years now!"

This story is one that Chuk would tell quite often: there were more than enough occasions where it fitted.

To end this, I shall relate another case, one that happened in real life (I was told this by E.L. Feinberg). Pomeranchuk was giving a talk at the Lebedev Institute (FIAN) on the diffractive production of particles in collisions of an incident particle (or nucleus) with a nucleus. In this process a frequent occurrence is that particles are produced when the incident particle (or nucleus) passes close to the nucleus, in the area of its diffraction shadow. L.I. Skobeltsyn was present at this talk. He asked: "How can this be, since the incident particle passes outside the nucleus and does not interact with it!" Pomeranchuk explained that the wave function of the incident particle overlaps the diffraction shadow of the nucleus and this leads to an interaction — and he continued his lecture. After a certain time Skobeltsyn repeated his question. Pomeranchuk gave the same answer but in different words and in greater detail. After a while Shobeltsyn asked the same question again. Pomeranchuk replied: "If you wish, you may consider this to be an immaculate conception".

3.3. A.I. Alikhanov

Physicist, citizen, director.

I consider A.I. Alikhanov to be one of my teachers (together with Landau and Pomeranchuk). He taught me much: a deep, non-formal

A.I. Alikhanov (Portrait by Bazheuk-Melik (1955))

understanding of physics, an ability to work with full dedication of self to the job, he taught me feelings of responsibility, courage, initiative, a civic spirit and civic courage, a genuine democratic approach (as opposed to sham democracy) and finally — he just taught me probity. He did not teach by edifying sayings, it is just that in any situation it was enough to imagine what his reaction would be and even what his thoughts would be on the subject, and it immediately became clear what one's behaviour should be, no other was possible.

Such behaviour was an intimate feature of his character and it manifested itself of course not only to me but also to anyone who had contact with him.

Alikhanov and Kurchatov were the founders of nuclear physics in the Soviet Union. These two candidates were considered when deciding who would be at the head of the nuclear program: they had been recommended by A.F. Ioffe. The choice of Kurchatov for this post was not decided on the basis of his higher scientific achievements (at the time Alikhanov already was a Corresponding Member of the Academy of Sciences and Kurchatov was not), but by the impression he made on Kaftanov, and later on Molotov. In the 1943 Academy elections (at which both Alikhanov and Kurchatov were elected to the Academy) one place was initially offered, and Alikhanov was elected to it. Only then was another place offered, and Kurchatov was elected to it. But generally speaking, one must say straightaway that as far as heading the program was concerned, Kurchatov was much better suited for the role.

A.I. Alikhanov was the founder and the first director of Laboratory No. 3 — TTL — ITEP. The Institute has been unusual from the very start. The director and his Deputy for Science, Vassili Vassilievich Vladimirsky, were not Party members and most of the heads of laboratories were not Party members either. Thanks to Alikhanov, the personnel of the Institute was of the highest caliber; their moral attitudes and scientific work were of the highest level. The Institute was created in December 1945 for the purpose of building heavy-water reactors. However, the very first governmental decree concerning the creation of Laboratory No. 3 included in the list of its tasks the study of high energy particle physics — which is the main area of study of ITEP today. This is proof of Alikhanov's brilliant foresight in science. Since reactors were needed, the existence of the Institute was tolerated, although it was always a great sore in the bosses' eyes.

Alikhanov did not like the Soviet regime. He understood the situation in the country clearly, he did not suffer from any illusions. In this respect he was fairly outspoken, in any case more outspoken than other prominent physicists of my acquaintance. In the 1950s he was in the habit of coming once or twice a week in the evening to the room where Rudik and I worked, and after discussing whatever was happening with reactors and after the question "What's new in theory matters?" he would move the conversation to general topics, often political ones. I have learned a great deal from these. Something that I particularly remember is Alikhanov telling us about the activities of Beria in Tbilisi, before he transferred to Moscow: he told us about people who were out of favour with Beria: how they would be seized in the street and tortured in prison, he told us how hunting chases were organised to get hold of women who were pleasing to Beria and whom he took as his mistresses. Their husbands were simply cleared out — either killed or imprisoned. I must stress that all this, including the general comment on Beria as "a dreadful person", was said before Beria's downfall.

One can add another facet to this illustration of Alikhanov's political attitude. He was the only eminent physicist who would visit P.L. Kapitsa after Kapitsa had been sent on Stalin's orders into exile to his country house near Moscow. Alikhanov continued these visits up to the moment when he was "summoned before higher authority" and told that unless he stopped his visits, he would be sent to the same place or even further. It was Alikhanov who told me that Kapitsa had been removed from his job and sent into exile because he had written a letter to Stalin saying that Beria was not competent in nuclear matters and could not head the atom project. Beria demanded a much harsher punishment for Kapitsa — his arrest with all the consequences this entailed — but Stalin showed mercy.

At the Institute, Alikhanov endeavoured to maintain an order where everything would serve the needs of science, while the myriad of bureaucratic and security restrictions would be minimized. This was not easy. The hierarchy of the Institute comprised the post of Plenipotentiary of the Defense Ministry (later of the Party Central Committee and the Council of Ministers). This post was occupied by Lieutenant-General (Ministry of State Security) Osetrov. His biography is remarkable: he had commanded the operation to evict from their land one of the peoples of Northern Caucasus. (This I heard from his adjutant who had taken part in this operation.) He also commanded the military division which guarded the site of the very first atomic bomb test. Osetrov had the right to act in some matters over the head of Alikhanov, but he understood that in case of a conflict with the director, one of them would have to leave and it was not clear who it would be. Therefore he preferred not to interfere in anything unless absolutely essential (unless he received an order from above). Therefore the Laboratory continued its existence as an island of freedom (a relative freedom, of course) and good sense.

TTL was also unique in its personnel. Alikhanov chose researchers only on the basis of their scientific qualifications (and of course of their probity, scoundrels were not taken on). The usual considerations of personal details — nationality, party membership — did not come into it. There were of course some difficulties, but Alikhanov succeeded in overcoming them every time. This concerned not only well-known scientists — well-known scientists with "bad personal data" could still be offered work in other places at least for a while — this also applied to young people, including engineers and technicians. The first interview with every future member of staff was conducted by Alikhanov himself. My own case may be quoted as an example. Of all the physicists graduating in 1949 from all departments of the Physics Faculty of Moscow University, I was the only Jew who was allocated a job in a good place. All the other Jews either were not given a job at all, spent a long time looking for something and eventually started to work in something completely alien (being a guide at the planetarium is a good example), or else they would be allocated to jobs in factories away from Moscow (this happened to Kirzhnits). I do not doubt that I owe my placement to Alikhanov and of course to Pomeranchuk who recommended me to him.

And finally, the administration and maintenance personnel of TTL. These were not numerous, the director chose them and controlled their work in such a way that it would serve science rather than their own interests, as usually happens nowadays.

Alikhanov devoted a great part of his life to the creation of heavy-water reactors. The first Soviet heavy-water research reactor was launched at TTL in 1949, i.e. only three years after TTL was organised. If you also take into account that the Laboratory was created from scratch and that the country had no experience whatsoever in creating heavy-water reactors (actually, the experience with graphite reactors was very modest too), this result is staggering. Less than two years after this, still under Alikhanov's leadership, an industrial heavy-water reactor was launched, producing plutonium and uranium-233. At the same time, still on Alikhanov's initiative, TTL began to work on the creation of heavy-water reactors for peaceful use — that is, reactors for nuclear power stations. One of these projects (this was the first reactor calculation that I carried out) was the project of a heavy-water fast neutron breeder reactor, working on the thorium-uranium-233 cycle. Work on this began in 1950. I would note here that this is the very cycle recommended in the 1980s by Nobel laureates Carlo Rubbia and Hans Bethe: they considered it to be the most promising way of nuclear power generation [41, 42].

The personal style of the Director led to the formation at the Institute in the 1950s of a working climate which was exceptionally encouraging for creativity. New ideas were boldly put forward, everyone endeavoured to work more and more productively, there was a constant exchange of ideas and proposals between colleagues, and personal relationships were very friendly.

All this resulted in a speedy development of young scientists at the Institute, so that they soon became independent. Here are a few examples taken from my own experience to illustrate what I am saying. I started to work at Laboratory No. 3 on 1st January 1950, after graduating from MSU. A.P. Rudik had started there almost at the same time (a few months earlier) and in the first few years we did most of our work together.

One of our main tasks in 1950–51 was to carry out calculations for nuclear reactors. We had no previous experience of this at all, so that at first we had not only to carry out calculations but also learn about them under the guidance of Pomeranchuk and Galanin.

Our experience grew and towards the end of 1950 — or beginning of 1951 — we already had a sufficiently good understanding of the physics of reactors to do calculations unsupervised, and even knew a few things about the fundamental problems of reactor physics. But we did not feel independent, we had an elder colleague above us who may not have had formal responsibility but who in fact was the one answerable for everything, including the reactor calculations done by us. This was Galanin, and higher up there was Pomeranchuk. We felt we were just carrying out orders, that although we were conscientious, we were run-of-the-mill. We did not feel compelled to take initiative.

The calculations we were carrying out were very important indeed and carried great responsibility: at the time the long-term, large-scale program of building nuclear reactors in the Soviet Union was under discussion, TTL and the Laboratory of Measuring Instruments of the USSR Academy of Sciences (LIPAN) were offering alternative proposals for this program. Alikhanov was the initiator of our Institute's proposals. He felt that due to their physical advantages heavy-water reactors were the most promising choice and that the emerging technical difficulties could be resolved by sufficiently nimble creative thinking. Since at that time the building of reactors was the main task of the Institute, Alikhanov kept himself fully informed of the way our theoretical calculations were going, he dropped in regularly every week (and sometimes more often) into our office (unless he was on a trip somewhere). He would discuss our results, compare our version of reactor parameters with that proposed by LIPAN, etc.

Some time early in 1951, when both Pomeranchuk and Galanin were on protracted business trips, Alikhanov called us and said that a letter from Zaveniagin had arrived at the Institute, demanding that the Institute present within two weeks its considerations concerning the construction of reactors. Since both Pomeranchuk and Galanin were absent, we were the ones who would write the letter presenting the Institute's proposals as well as details of the parameters for reactors. We were seriously frightened — in 1951, writing such a letter to Zaveniagin "himself" was definitely not a joke. But we had no choice. We were scared stiff, but we did write this letter, after checking all our calculations. Alikhanov signed this letter and it was sent off. From that moment we became independent and were no longer afraid to assume responsibility.

That was the style of Abram Isaakovich: he tried to have personal contact with every member of staff independent of his rank (and that difference for us was huge: Abram Isaakovich was an Academician and we were junior research scientists with just over a year of work experience). Such intercourse, always in an informal setting, gave Abram Isaakovich a personal impression of the ability and qualification of the researcher, how conscientious he was in his work, and when that impression was positive, he would begin to treat this person with confidence. Naturally, such attitude would inspire that person to work even better.

And another example — less important but relevant.

Sometime in 1951 or 52 Alikhanov called us — Galanin, Rudik and me — and asked us to write a report on a classified document. The name of the author was unknown to us; the content of the document was an explanation of the structure of atomic nuclei. The document came with a large box containing meticulously crafted wooden objects which were meant to be assembled in a certain order to form nuclei according to the author's theory. But the main issue in all this was the first page which had a handwritten message: "To Academician A.N. Nesmeyanov. Please assess"; signed: Beria. This was followed by a formal message from Nesmeyanov (President of the Academy) to Alikhanov. Our Director, aware of our feelings, said: "Write what you think. I shall sign it, it will be sent with MY signature". After this we had no difficulty in writing our comment. It was sent off — and nothing happened. Much later I heard that the author had been the Head of the Kolyma GULAG camps. That explained everything, from Beria's signature to the high quality of the wooden parts.

It was mostly due to its Director that ITEP was an utterly unique scientific institution in the 1950s. I do not know of any other similar institute, and it may well be that there was indeed no other in the USSR. Everything in ITEP was subordinate to one aim: to science, pure or applied. And one thing was valued in science — the end result. Each and any scientist might come on any day to see the Director, who always found time to discuss science with him, and — I emphasize — this was not a casual brief conversation, but a businesslike discussion, in which all details were discussed. If Alikhanov could not talk to his member of staff during the working day[15] he would invite him to come in the evening, after 6 or 7 pm. But he never postponed the discussion for long. New ideas in science were particularly valued and if they concerned experiments, they definitely took pride of precedence. If Alikhanov decided that the new idea was indeed significant, he would take it on with enthusiasm and he would infect others with this enthusiasm, so that work on it would start immediately. As a result, it is in ITEP that very many ideas for experiments and methodology were initiated. This was the case for creating a strong focusing accelerator, for bubble chambers, for working out experiments on parity violation.

Subsidiary services at the Institute had to work exclusively for matters of science. Alikhanov did not allow their excessive growth, as he understood perfectly that in such a case they would begin to work for their own benefit and even to hamper scientific activity. For instance, in the early 50s, when the Institute was no longer so tiny and when much had already been achieved, the personnel department and the secretariat together consisted of one man, who also would type out all necessary documents himself. Alikhanov demanded that subsidiary personnel (just like scientists) work with energy and initiative, that their work be concrete, and if this did not happen, he would tear them to pieces. From time to time one would hear something like this issuing from his office: "How dare you work like that — you deserve your balls to be torn off and thrown out into the street. The dogs would come, sniff at them and wouldn't eat them!" And as a rule this sort of telling-off made a man understand that he should do better work, otherwise he would have to leave the Institute. Any reasonably competent worker would not wish to leave — it was a pleasure to work at our Institute.

---

[15]In his position Alikhanov was forced to deal with administrative questions which he intensely disliked. "After this sort of business one's head feels like a head of cabbage" he used to say.

Everything about the Institute interested our Director. Science was of course his first priority, but everything else was seen by him, nothing escaped his attention: starting with the program of a seminar and the condition of the library and ending with a broken or dirty toilet. Whenever he saw the least failure in order, he would react immediately: he would call out the guilty person, he would demand that the defect be rectified immediately. Woe to him who tried to hide behind "objective reasons". Alikhanov knew himself how to put things right, therefore it was difficult to contradict him. There was an incident when he personally undertook to adjust the working of our drains, and not any old drains — this was a section that had to function not from the top down but from the bottom up — and he did get it to work properly.

Alikhanov knew and loved music, he was on friendly terms with Shostakovich, he often went to concerts of classical music. He said once to Shostakovich who had come to visit — "You have a wonderful home, but I cannot understand how you can live so far from the Conservatoire". Alikhanov's wife, Slava Solomonovna (née Roshal') was a violinist, in 1935 she got a prize at the "Young Talents" competition. His son Tigran was a pianist and was for several years the Director of the Moscow Conservatoire.

In 1951, however, the successful creation and development of the Institute came under serious threat. The causes of this were political. As I already said, the Technical Laboratory seriously irritated the powers-that-be. And a PGU Commission was sent to the Institute to check on it.[16] At that time Alikhanov and his deputy Vladimirsky were at our test site preparing the launch of a reactor, and the duties of Director were carried out by Sergei Yakovlevich Nikitin (by the way, he too was not a Party member). The aims of the commission were obvious — it was collecting incriminating material ("kompromat" in Russian). The commission studied documents and interrogated all personnel. The questions were varied, and very frequently they were provocations — constructed so that they would

---

[16]PGU is the First Main Directorate of the USSR Council of Ministers which was dealing with the atom project. In 1946–53 its Head was B.L. Vannikov, and A.P. Zaveniagin was his deputy.

provoke a compromising answer. For instance, I was asked to name the last book I had read and stupidly I said "A Balzac novel". I later heard that I was therefore found guilty of reading bourgeois literature. I was also asked how many projects I had done while working at the Institute. There were 11 projects, of which 6 were secret and 5 could be talked about. All of them had been carried out in collaboration with A.P. Rudik. As I later heard from Nikitin, who as Acting Director was a member of the Commission, that once I had left the room, the chairman of the Commission, a Colonel of the Ministry of State Security, proposed to fire one of us — me — and to give only secret assignments to the other — Rudik. Nikitin had great difficulty in keeping me at the Institute. He argued that there were more secret projects than non-secret ones and also that the work was done faster and better when two people were working together in cooperation. The members of the Commission only stepped back when Nikitin asked whether they would take upon themselves all responsibility in case the firing of one of the theorists would result in failure to complete the jobs set by assignments of secret work.

But the results of interviews were not always as happy in some other cases. Basing himself on the work of this commission, Zaveniagin signed a decree which would ravage the Institute and just about destroy its work. Several dozens of the best researchers (mostly, but not exclusively, Jewish) had to be dismissed, the Director was accused of serious financial and practical infractions — he was even accused of criminal activity. (For instance, it was stated that one of the cottage-type houses built by the Institute had been stolen.) There was also an attack on Pomeranchuk personally: a separate paragraph declared that Pomeranchuk was an "inveterate holder of multiple jobs".[17]

---

[17] A year or two before this, Pomeranchuk had been appointed, again by decree of the selfsame PGU, to the post of part-time head of the Theory Division in the Hydrotechnical Laboratory (now JINR, Dubna). He regularly went to Dubna once a week, he actually created the Theory Division by sending several of his pupils to work there, he held many discussions with experimentalists, directing them to tackle genuine problems. He did not take any payment for this work — not one penny — although Dubna tried very hard to pay him. Therefore it was not difficult to remove this point at the time when the original text was being changed into the final version of the decree (signed by Vannikov), but Pomeranchuk was forced to leave the post of Head of the Dubna Theory Division.

And here S.Ya. Nikitin did something unheard of at the time: he refused to carry out the decree! He declared that he could not carry out the decree in the absence of the Director, and he managed to maintain the situation in this state for a month or two. During this period the reactor was successfully launched at the test site. Alikhanov came back in triumph, he went to Vannikov and succeeded in having the decree annulled — or rather modified. In the new decree fewer people were dismissed — 10 or 12 — (but these still were very good at their work and they were all Jewish), accusations of financial crimes also fell away, the Institute survived but sustained heavy losses. Nikitin's (impertinent) act was not forgiven: a year later he was removed from his post of head of laboratory and demoted to a non-managerial post, under an entirely inconsequential pretext. Alikhanov succeeded in reinstating him only two years later.

A few years after that, there was an attempt to remove the Director. A certain Romanov was appointed Secretary of the ITEP Party Committee. Soon after his appointment he started a campaign aimed at having the Director removed: he wrote denunciations, etc. He even achieved some success, that is some support from higher on. But he "collapsed" in the most stupid way. There was a certain lady working at ITEP as a flower woman. She looked after the flowers that grew in decorative beds and in the winter garden. (At that time ITEP had the most beautiful flowers, they were well cared for, there was even an indoor "winter garden".) The lady in question was not very strict in her behaviour and enjoyed some success. She lived in one of the cottage-style houses that belonged to the Institute, she had a room in a communal flat. Romanov began to court her, and this was immediately noticed by her neighbours (who shared the flat). They established a correlation of events: if there was intense culinary activity in the kitchen of an evening, Romanov would soon appear. As soon as the door of the lady's room would close, the neighbours would immediately take turns in putting an eye to the keyhole in that door. They were indignant at the happenings and wrote therefore to higher authority and also informed Romanov's wife of them. There was an investigation of the case and Romanov was dismissed from his post "for amoral behaviour".

An ever greater danger threatened ITEP (TTL) in 1956, when the Secretariat of the Central Committee of the Communist Party of The Soviet Union decided to disband the Party organisation of TTL: many of its members were expelled from the Party and four were fired from their jobs. The events of 1956 and the role Alikhanov played in them — he literally saved the Institute — are described in Yu.F. Orlov's book "Dangerous Thoughts" (Moscow, AiF, 1992). After the 22nd Congress of the Soviet Communist Party, at which Khrushchev spoke about Stalin's personality cult in his key-note speech, every Party organisation in the country held meetings at which Khrushchev's speech was discussed. At the ITEP meeting, four participants — Orlov in particular — made speeches with appeals for democracy to be established in the USSR. (Some of these speeches were rather confused. For instance, in one of them, Analov made an appeal "to arm the people".)

I am quoting excerpts from the resolution of the Central Committee which Orlov did not know about. The session was held on 3rd April 1956, chaired by Suslov and attended by Central Committee Secretaries Beliaev, Brezhnev, Pospelov, Furtseva, Shepilov, also several Members of the Central Committee, and others. The resolution was headed "On hostile sorties at the meeting of the party organisation of the USSR Academy of Sciences' Thermotechnical Laboratory against the conclusions of the 20th Congress of the CPSU". The resolution said that at this party meeting "anti-party statements were made by junior members of science staff Avalov, Orlov, Nesterov and technician Shchedrin, in which they made libellous vicious, provocative declarations, aimed at revising the general line of the Communist Party ...". It was noted later that "In TTL... an unhealthy, rotten situation had arisen (particularly amongst members of the Communist Party in the science sectors)". A decision was formulated to be ratified by the CC Presidium: — I quote:

"The CC decrees:
1. To approve the decision of the Political Administration of the USSR Ministry of Medium Machine Building to expel Avalov, Orlov, Nesterov and

Shchedrin because of their hostile, anti-Party and anti-Soviet speeches at the Party meeting in TTL AN USSR.

2. To state that the Party organisation of TTL AN USSR turned out to be politically unsound and lacking in steadfastness. In view of this, to instruct the CPSU Lenin regional committee of Moscow, together with the Political Administration of the USSR Ministry of Medium Machine Building, to revise the registration of CPSU members and candidates to membership at TTL AN USSR, with the view of keeping within the ranks of the Party only those who are indeed capable of applying the Party's general line.

3. To put the re-created Party organisation at TTL AN USSR under the authority of the CPSU Lenin regional committee of Moscow.

4. To remove from his post the head of the political department at TTL AN USSR, comrade Shmelev I.S., for the reason of his not coping with the task entrusted to him.

5. To note that the Political Administration of the USSR Ministry of Medium Machine Building (comrade Mezentsev) had not maintained the necessary surveillance and control over the work of the Party organisation and had (repeatedly) not noticed major defects in the choice and education of cadres carried out by the administration and the political department of TTL.

6. To oblige the leadership of the USSR Ministry of Medium Machine Building (comrades Zaveniagin, Mezentsev) to take measures for adding strength to TTL AN USSR by appointing leaders for the scientific, engineering and technical work.

Point 6 of his decree was particularly dangerous: it threatened mass purges at TTL. Alikhanov saved the Institute yet again. As he told me, on the next day after the Party meeting (or rather, after its second day, since it lasted for two days) he was handed an order from State Security to recall permission to enter the Institute's territory from Avalov, Orlov, Nesterov and Shchedrin. In this case the Director could do anything — he was obliged to demand that the people in question immediately hand in their passes. At this very moment Alikhanov took the internal telephone which maintained a direct link with the Kremlin and called Khrushchev directly. In his conversation with Khrushchev (although the leader, as Alikhanov said, was obviously furious) the Director succeeded in achieving much success. He was promised that the Institute would continue to function, that there would be no more dismissals and, moreover, that the "measures for adding strength" to the science personnel would be carried out with his — Alikhanov's — knowledge and approval. But his attempt to save the four men was a failure. Alikhanov tried to defend them by saying "But these are boys ...". Khrushchev retorted sharply: "These boys were attacking the foundations of the state and they will be severely punished!"

The problem of "strengthening leading personnel" was solved by Alikhanov in the best possible manner: he invited M.S. Kozodaiev to become Assistant Director. This was a member of the Party, but he had worked with Alikhanov in the old days (at the Leningrad Physical-Technical Institute), he was a very decent human being.

One of Alikhanov's main achievements was the creation in the USSR of strongly focussed high energy proton accelerators. It is known that the idea of strongly focussed accelerators came from the US, but it was immediately grasped by V.V. Vladimirsky at TTL, and under his leadership the Laboratory created first the project of a 7 GeV and later of a 50–60 GeV one, which at the time was the biggest in the world. As far as this second is concerned, Yu.F. Orlov and D.G. Koshkarev played a major role in its development. (Koshkarev found a way of passing through the critical energy; this was not known in the US at the time.) Alikhanov loved the idea of building strongly focussed accelerators and began to turn it into reality with his typical energy. He succeeded

in the TTL being granted an adjoining piece of land and this is where
the construction of the 7 GeV accelerator began. He encouraged and
organised all experimental groups to work on the future accelerator,
he pushed through both paperwork on the project and the actual
construction work. There were no serious objections to creating the
7 GeV accelerator, but the proposal to build a 70 GeV accelerator
met with great opposition. Against it stood the so-called "4Bs" as
they were nicknamed in TTL — Bogoliubov, Blokhintsev, Burlakov
(who at the time was a leading light at the department of the Central Committee in charge of atomic energy) and also B.L. Vannikov.
The main argument of opponents of the accelerator was — "How can
ITEP, a comparatively small Institute, build the biggest accelerator
in the world?" Alikhanov retorted to this argument by "But after
all we do know of cases when a weak, fragile woman gave birth to
an exceptionally strong man!" With Kurchatov's support, Alikhanov successfully overcame this opposition, and a decision was taken
to build near Serpukhov a 70 GeV proton accelerator according to
the ITEP design and also as a daughter establishment of ITEP. At
a later time, Bogoliubov's group changed its position, attempted to
take over the future accelerator and succeeded in this. In his fight
against this turn of events, Alikhanov had a heart attack — right in
the office of Petrosiants, the chairman of the Committee on Atomic
Energy.

Unfortunately, Alikhanov also made mistakes. The main one —
also the saddest one — is the story about the discovery of varitrons, particles whose mass was meant to be intermediate between
the masses of muons and protons. Alikhanian and Alikhanov with
their staff (the main part here was played by Alikhanian, since Alikhanov concentrated on reactors) built a magnificent appliance: a
magnetic spectrometer. This was a large electromagnet which had
lines of counters placed between its poles. This magnetic spectrometer allowed to determine with great precision the momentum of
a charged particle at its entrance into the spectrometer. In order to
determine the mass of the particle, one needed to know one more
characteristic — its energy. The energy of a particle was determined
by its ionisation trajectory in the filters which the particle hit after

going through the spectrometer. One of these appliances was placed at a cosmic ray station on mount Ararat (3200 metres) in Armenia, and another, of a smaller size, was stationed in ITEP. The mass spectrum of cosmic rays obtained by using the magnetic spectrometer on Mount Ararat showed a great many peaks which were interpreted as being hitherto unknown mesons and received the name of varitrons. (The data obtained on the spectrometer stationed at ITEP, that is at sea level, were less defined. This spectrometer was used mostly for checking the methodology.)

The experiments of Alikhanov, Alikhanian and their staff were strongly criticised by physicists working at FIAN: Vernov, Dobrotin, Zatsepin: the very existence of varitrons was subjected to doubt. Additional research showed that the criticism was justified: varitrons do not exist. The mistake of Alikhanov's and Alikhanian's groups was due to their measurement of energy by the range of particles in the filters. It was supposed that the losses of energy were only due to ionisation. But actually a particle loses a considerable fraction of its energy as a result of meson production and of inelastic collisions with nuclei, i.e. not by ionization.

To be fair, one must note that part of the responsibility for this mistake lies also with the theorists, especially Landau and Pomeranchuk: as their work developed, Alikhanov and Alikhanian discussed it many times with them. The fact that Landau missed this mistake which may seem trivial (at his level) can be understood if one takes into account his inner convictions: Landau did not believe in meson theories, and the fact that a great many mesons appeared to have been found showed from his point of view that meson theories had nothing to do with real physics.

Alikhanov made another mistake in 1962, when he gave support to experiments of Ya. Shalamov and their theoretical interpretation by A. Grashin. Grashin and Shalamov were stating that they had discovered ρ mesons. These experiments were energetically criticised by a great many experimentalists and theorists at ITEP and it was shown that one could not make any deductions from these experiments. (The author also made his contribution to these critical remarks.) In answer to scientific criticism, Grashin moved the

discussion into another area — that of political accusations and ca-
lumny. Here Alikhanov immediately withdrew his support (he could
not stand this sort of thing), and Grashin was dismissed from ITEP.

In praise to Alikhanov one must say that he never took any administrative measures against staff who criticised him. On the contrary,
he gave N.G. Birger a job, though she had expressed criticism of his
and Alikhanian's work when she was working at FIAN. He gave her a
job at ITEP when she was dismissed from FIAN on the grounds of her
"bad" nationality. The same N.G. Birger also criticised the work of
Grashin and Shalamov, but Alikhanov's attitude towards her did not
change. And lastly, I, together with other theorists, wrote a special
paper (never published) which contained mathematical proof of the
fact that the experimental data obtained by Shalamov and Grashin
could serve for whatever deductions one chose, in other words the
two had not made any discovery at all. This conclusion was of course
unwelcome for Alikhanov who wanted outstanding discoveries to be
made by his Institute. This, however, did not in any way affect our
relationship — this remained very warm right to the end of his life.

Alikhanov cared not only about the workplace duties of his staff,
but also about their private needs. As soon as he heard about some
difficulties or problems — whether they had to do with health, accommodation or even family — he would readily try and help and
he needed no reminders for this. I could relate several such cases,
but I shall tell about just one because it concerned me personally.

At the end of 1950s I put to the Directorate a request to be allocated a flat. At that particular time, the building of a block of flats
for institute staff was nearing completion and I had been allocated a
flat in it. However, the local area's Party committee did not ratify
this allocation, for its own formal reasons. I had no success with
my appeal to the local authorities (Mosgorispolkom) either. Then
Alikhanov decided to make a personal visit to the vice chairman of
Mosgorispolkom, the man who was the principal authority in matters of allocating accommodation in Moscow. I met Alikhanov in the
hall of the Institute on his return. He was just getting out of his car.
On his jacket he was wearing the Golden Star of a Hero of Socialist
Labour, which he wore very rarely indeed. He was very upset and

said, pointing at the Star: "Look, I even put this on for your sake, but it did not help". This sentence almost compensated the loss of the flat for me.

I have already said that Alikhanov would regularly (sometimes several times a week) come to the office which I shared with Aleksei Petrovich Rudik. This went on until the moment when he fell seriously ill. He would come towards evening as a rule, but sometimes he came in daytime. In the latter case, if his secretary came during our conversation and told him that he was wanted on the telephone on some important administrative business, he would answer as a rule: "Let them phone in an hour's time. Right now I am busy". For him, it was more important to talk with theorists than to tackle administrative business. If there was something concerning reactors, our conversation would start with them. He often initiated a discussion on some problem connected to experiments currently carried out in ITEP or possibly to the latest experimental outside news. And always, practically every time he came, he asked at some moment: "What's new in theory?" It was not easy to answer this question, because one could see in Alikhanov's reaction that he was truly interested to know the news happening in theory, so that a formal answer was no good. One had to answer in such a way that he would understand, but it was not possible to use the mathematical techniques of theory because this was beyond the limit of his own knowledge. Therefore one had to seek physical explanations, which was difficult but fascinating. As a result we would have a lively discussion about physics which would give us a great deal of pleasure (and presumably it gave Alikhanov pleasure too, otherwise he would not come to us so very often).

3.4. A.I. Alikhanian

Sketches for a portrait against the background of the times.

Artemii Isaakovich Alikhanian was one of the founders of nuclear physics and elementary particle physics in the USSR. He was one of the first in the country to understand that it was absolutely

A.I. Alikhanian

necessary to carry out experiments with high energy particles in order to reveal the nature of nuclear forces (that was the formula used at the time), i.e. to develop particle physics. This in its turn necessitated a close cooperation between experimentalists and theorists. Experimentalists must know what is happening in theory research and listen to the opinion of theorists — they may not need to follow their suggestions, but they must listen and be aware. Theorists must also know about what is happening in experimental work. In accordance with this idea, Alikhanian initiated (starting in 1957) a series of conferences and schools on the physics of elementary particles and of high energies in Yerevan, and later at the Nor-Amberd cosmic ray station. This was very timely.

Conferences and schools of this type had not been held in the USSR for nearly 20 years. Over this time, science had made immense advances, many young scientists had joined the ranks. However, those young people who had come to nuclear physics (as well as those of an older generation) worked mostly on atomic problems — those concerning the physics of nuclear reactors and of the atomic bomb. Both theorists and experimentalists were short of knowledge about the latest development of elementary particle physics. This gap was filled by the conferences, and later also by the schools which were organised by Alikhanian. This task was accomplished with his characteristic brilliance and organizational talent.

The plane carrying the participants to the first conference — at the time the Yerevan route was served by small IL-14 type planes — held the flower of Soviet physics: Migdal, Zel'dovich, Ginzburg, Feinberg, Berestetsky, Pontecorvo, Chudakov, some others whose names my memory has not retained, and also many physicists of the young generation. Alikhanian met us at the airport, right at the foot of the plane's gangway. We were taken to our hotel (Yerevan's finest at the time), and an hour later was the start of what in those times was impressively named "a banquet" and which now is called a "welcome party". I remember at the banquet (1957 was the time of Khrushchev's political "thaw") E.L. Feinberg proposed a toast "To the end of the problems between Armenia and FIAN!" (Up to that time there were difficulties in the relationship between the Yerevan group — and primarily Alikhanian himself and also Alikhanov — and the FIAN group — Vernov, Dobrotin and Zatsepin: research on varitrons was subject to fierce criticism. After 1957 there was a détente: it became clear that the varitron work was a mistake. Alikhanov did admit implicitly that varitrons did not exist, and relations returned to normal.

Then the conference started. Its program had been worked out by Alikhanian — it was an excellent program, each participant shared everything he knew with the others and also his own knowledge was enriched.

Alikhanian was aware that man does not live by bread alone. He organised a magnificent cultural program: trips to Garni, Geghard, Etchmiadzin, Jermuk. At that time the temple in Garni still lay in ruins, only the columns stood upright, the capitols were scattered on the ground all over the place. Nonetheless, the temple produced the strongest impression: one only had to step up to the edge of the precipice above which the temple stood, a sheer, almost vertical drop of 300 or 500 meters, and to imagine how strangers, wandering savages, would emerge at the other side of the valley and would see the temple towering above. Zel'dovich wanted to come to the edge of this abyss as well, but he was stopped when the security man who accompanied him shouted: "Yakov Davidovich, step away from the edge!" In front of the temple there is a flat piece of land and there, shadowed by trees, stood tables with a delicious treat which also included a new wine, "Vernashen" that had only just appeared at the time. Around these tables, in such an inspiring setting, we would talk about science and new ideas were born.

Another memorable trip — to Jermuk. We went there by coach and we talked science again right through the long voyage. Our route lay through the frontier zone. A frontier guard came into the coach and began to check our passports. As he made his way along the list I quoted earlier, his face got darker and darker. As he came to the name Chudakov, he said: "Ah, at last we get at least one of the right kind!"

After visiting Jermuk, Berestetsky, Vaisenberg, Goldman and I decided to walk to Sevan, crossing the Vardeniz mountain ridge on foot. We were lightly clad: Vladimir Borisovich Berestetsky had flimsy slippers and pyjama bottoms on, the others also wore the sort of footwear that is usual among tourists. I was the only one to wear American army boots, of the kind nicknamed "Studebekkers". To start with, everything was wonderful — it was spring, the month was May, flowers were blooming, birds were singing. Later a few snowflakes appeared, then snow covered the ground, then the snow came up to our knees. I walked at the head of our procession, opening the way with my boots, the others followed literally in my footsteps.

We came to a mountain stream which was running through snow-covered banks. It had to be forded. We could see that this would give us no pleasure: the water temperature was close to freezing and we would be emerging onto a snow-covered field where we would not even be able to dry our feet. I kept insisting that we ought to ford that stream without delay, since there was no other choice. If we crossed that stream, we still had a chance of getting out of the snow-fields and even maybe to reach Sevan. And here I saw that Vaisenberg was walking upstream along the river. There, at the top, the stream was covered with a snow bridge, and water was noisily gushing from underneath that snow bridge, making a fantastic noise. I realised that Vaisenberg was meaning to cross the stream using the snow bridge. This was appallingly dangerous: that snow bridge was obviously already weakened by the stream, one could be almost sure that it would collapse under a man's weight, and Vaisenberg was unlikely to survive. I shouted: "Aleksandr Ovseevich, stop, come back!" He walked on. I shouted again — no effect. Only a few meters to go before the snow bridge. And then I collected all my (restricted) knowledge of foul language and poured it all over him. That worked. Vaisenberg turned back. I reckon that I saved his life.

We crossed the stream using the Tadjik method, that is holding each other tight in an embrace. After that stream, the snow fields became less frequent and by evening we reached a dry, completely bare space. We found a hole in the ground, crawled into it and huddled close together. We had very little food, but we did have a bottle of brandy. We drank the lot, which made us feel a bit warmer, and we stayed there till daybreak. In the morning I turned out to

have developed snow blindness: I could not open my eyes, the pain
was horrific. We set off again, I was holding onto Goldman, just
like a blind man holds onto his carer. We discovered that we had
actually almost reached the road — a bad road, obviously barely
used, but a road nonetheless. But then the last obstacle appeared —
the road was blocked by a snowdrift at least 10 meters wide. One
needed to traverse it one by one. This was most difficult for me —
to keep my eyes open, in spite of the strong pain and the streaming
tears. On the following day we reached Basarguechar and contacted
Alikhanian on the telephone. He asked: "Where are you now?" —
"In Basarguechar". — "How on earth did you get there?!" A few
hours later a car arrived to take us back. Later, once we had returned
to Moscow, Vaisenberg told me that he had never in his life felt better
than he did after that excursion.

Alikhanian had told me that the most interesting place in
Armenia was Zanguezur. The following year, four of us decided to
visit there — Abrikosov, Goldman, Sudakov and I. We decided that
we would take a plane to Kapan and from there we would walk to
Tatev, and Alikhanian promised me to send a car to Tatev to collect
us. Tatev is one of the most remarkable places in Armenia, it had
the very first university on the whole territory of modern USSR, they
taught mathematics there as early as the XIII century! Everything
was already arranged, but suddenly Volodia Sudakov turned up and
said that Marina, a pupil of Alikhanian (she later became his wife)
wanted to come with us, that he — Sudakov — had invited her. I did
not like this at all. I went to the transport office to finalise arrange-
ments about the car and the car turned out to have broken down.
Then I went to Marina and described to her all the difficulties that
lay ahead: our climb would be long, steep and difficult, the path was
bad, we might have rain and this would make everything slippery,
therefore our progress would be even more difficult. As a result, I
managed to talk her into giving up her plans to come with us. The
car was mended straight away, and we set off to our plane.

We were indeed unlucky with the weather and by evening, when
we arrived in Tatev, we were wet through, covered in dirt and hungry.
I went to the village administration (the only place with a telephone)

to call Alikhanian, my friends stayed outside. When I came out, they were already surrounded by a crowd of curious villagers. At that time, very few strangers ever came to Tatev, the appearance of anyone new was a real event — and here they had not one, but four Russians coming to the village. We didn't quite know what to do: night was approaching, rain was falling, we did not have a tent and we did not know where and how we would spend the night — Tatev was just a large village and there obviously was no hotel or guesthouse. Moreover, no-one around us spoke Russian. But then a young guy came to us and in broken Russian invited us to spend the night in his home. We were delighted to accept. On the way to his house he explained that he had done his military service and therefore knew some Russian. He was living with his sister, their house was poor but they would do everything they could for us. We arrived at the house. It was indeed poor: one room, a table, benches, two beds with basic mattresses. It was however warm and dry. The guy explained that we would sleep there, and that he — together with his sister — would find somewhere else to spend the night. Here the sister appeared — a very charming girl about eighteen years old. She carried a bowl containing warm water. The guy explained that according to an ancient Armenian custom a woman must wash the feet of the tired stranger who comes to the house. We didn't really know what to do: our feet were dirty, the girl was ravishing .... But one may not insult the master of the house. We tried very hard and used all our diplomatic skills to persuade our hosts to forgo their ancient custom. Then they brought us food. It was scant, but one could see that this was all they had in the house. And then they left, leaving us to rest.

Our car arrived the following morning. We had used our time to visit the IX century Tatev cathedral and the XIII century monastery, built on a cliff 400 meters above the river (the university was housed in the monastery too). The courtyard of the cathedral had an amazing monument — an oscillating octagonal stone pillar which stands on a hinged base, 8 meters high. You just touch it, and the pillar immediately starts to oscillate.

The time had come to part from our hospitable friends. There could be no question of payment for the overnight stay — this would cause mortal offence. The young man unexpectedly helped us in our dilemma — he asked us to take a picture or two of him with his sister and to send the photographs to them later. Abrikosov was the only one to have a camera, he immediately fulfilled this request and took their address. Later, in Moscow, I reminded Abrikosov many times that we had to send those pictures to Tatev. At first he would say that the film had not been processed yet, then he declared that it was the usual thing when one meets people at a hike to promise them to send photographs, and then not to send any. At this I asked him to give me both the film and the address, but he said that he had lost them. I could never forgive him for this. I shall always remember these young Armenians who brought intact to our times the patriarchal way of life of ancient Armenia.

This whole atmosphere on the one hand favoured work and creativity, and on the other allowed one to relax in the company of friends and colleagues — this atmosphere reigned at all conferences and schools organised by Alikhanian. Later these schools were held at the cosmic ray station in Nor-Amberd. It was a great pleasure to ride up on one's skis from the lower station to the higher station (pulled by a tractor) and after a few days working at the top to ski downhill under one's own steam.

Although all schools have been interesting and useful (I still use the lectures presented at them), but the one held in 1965 was the most interesting one. Alikhanian had been able to invite M. Gell-Mann, L. Lederman, T.D. Lee, M. Schwartz — laureates and future Nobel laureates, also S. Goldhaber, M. Strauch and several more prominent physicists from abroad, and he also succeeded in including Pomeranchuk in the group of Soviet physicists — though Pomeranchuk only rarely attended schools or conferences. This school became a major event in our life. Of course, the fact that such prominent physicists had agreed to come to Yerevan was due not only to the stature and charm of Alikhanian, but it was also in part due to the Yerevan Physics Institute (YerPhI) having built an electron ring accelerator (EKU).

The creation of EKU, the creation of a cohesive group of physicists

working on it, the working out of programs for experiments on EKU
and the implementation of this program — all this is incontrovertibly
an achievement of Alikhanian's. He took extraordinary decisions concerning the complementation of staff. He appointed Yu. F. Orlov to
the post of chief theorist for carrying out calculations of EKU: Orlov was a well-known dissident and human rights activist, previously
employed by ITEP, who in 1956 had been thrown out of ITEP and
expelled from the Party by decision of the Secretariat of the CPSU
Central Committee. Alikhanian was fully aware that he ran serious
risks by offering a job to Orlov: YerPhI was part of the same Ministry of Medium Machine Building as ITEP was, and before he was
employed at YerPhI, Orlov could not get any job anywhere at all —
he was a marked man. Alikhanian was influenced not only by his
professional interests, but also by moral considerations when employing Orlov — he also wanted to support him. Later he succeeded
in pushing through the election of Orlov (for his great contribution
to EKU) to the Armenian Academy of Sciences, as a Corresponding
Member. (This status proved very useful for Orlov: the money that
he received as an Academician was for a time his only source of in-
come and allowed him to survive at a later time, when persecutions
against him were resumed.)

Another similar case (though of course a less serious one) concerns
the decision to offer a job at YerPhI to the theoretical physicist V.A.
Khoze who until then had been a post-graduate at the Institute of
Nuclear Physics of the Siberian Branch of the Academy of Sciences
(INP SO). Khoze was a member of the local "Voluntary People's
Patrol" at the Novosibirsk "Akademgorodok". One evening, being
on patrol duty, Khoze saw a young man creating a rowdy scandal
in a restaurant and demanded that the young man follow him to
the police station. The young man, however, declared that Khoze
would bitterly regret this, as he happened to be the son-in-law of
Academician Lavrent'ev, Chairman of the Siberian branch of the
USSR Academy of Sciences. The young man's wife, who was present,
also rose in defence of her husband. Khoze nonetheless brought the
VIP son-in-law to the police station. Trouble then started. The son-
in-law escaped unscathed, of course, but Lavrent'ev began to demand

that Budker, the Director of INP SO, dismiss Khoze's from his post-graduate position. The institute's theorists (their majority) did not want to give Khoze away, since he was a good postgrad and anyway they considered that he had been right in that incident. Budker stood firm for some time, but then he called the theorists and said: "Your Khoze has already cost me 5 million roubles. I can no longer put the Institute at risk because of him". Then the theorists turned to Alikhanian, who offered Khoze a job, although this was risky for him as well: Lavrent'ev was powerful not only in Novosibirsk, but also in the whole Academy, and he never forgot a grudge: he kept persecuting Khoze's parents for many more years. Just as in the case with Orlov, Alikhanian did not have to regret his decision: Khoze did much good work at YerPhI.

The last International School on Theoretical and Experimental Physics organised by Alikhanian ran in Yerevan in 1971, from 23 November to 4 December. A paper by Alikhanian and Orlov was presented at this school, on the project of an electron-positron collider of 100 GeV center-of-mass energy. This project was very similar to that of the future LEP collider in CERN (there were no plans for LEP at all at that time, though!). The Alikhanian and Orlov project was not implemented and it was not even published. This was due to the fact that after the invasion of Czechoslovakia and protests from dissidents, pressure on them began to intensify. Orlov was still considered to be a dissident, also he refused to apply for re-instatement in the Party and he openly condemned the invasion. Alikhanian came under pressure to remove Orlov from the list of the project's authors. He refused, and the project was axed.

At the 1971 Yerevan school, Khoze and I presented a paper on a proposed experimental program on the future e^+e^- collider with beam energies of 50 to 100 GeV. In fact, this could have been a program for the future LEP collider. Some parts were missing — experiments with charmed particles, which had not yet been discovered, and experiments for the production of the Z^0 boson, in whose existence few people believed at that time (neutral currents were discovered in 1973), but in all other respects, including experiments of $W^\pm$ production and the measurement of the cross section of $e^+e^- \rightarrow$ hadrons, it was the LEP program. The proceedings of the Yerevan School were never published — for the same reason: the seditious lecture of Alikhanian and Orlov. Our lecture also remained only as a preprint of the Yerevan Institute EFI-TF4 (1972), which contained a reference to the lecture of Alikhanian and Orlov.

If the Alikhanian-Orlov project had been implemented, the world centre of high energy physics would have moved to the USSR. But this did not happen, politics succeeded yet again in throttling science.

In 1976, the Institute of High Energy Physics (IHEP) put forward a proposal to build at that Institute a proton accelerator with superconducting magnets and with a beam energy of 2 TeV. This proposal had most energetic support from A.A. Logunov, who was at the time head of science of IHEP, Vice-President of the USSR Academy of Sciences, Rector of Moscow University, member of the CPSU Central Committee from 1978, etc. An extended council meeting of the IHEP Science Coordinating Committee was called in Protvino to discuss this topic, and physicists from several institutes were invited to take part. The aim of this event was clear: it had to approve the IHEP proposal to build a proton accelerator. Therefore all putative participants prepared papers which argued (from various points of view) that such an accelerator indeed needed to be built and put forward a program of various experiments to be carried out on it.

I too was invited to take part in this meeting. I was however still of the opinion that the most sensible action, which promised important results in the nearest future and which was also entirely realistic, was the building of an electron-positron collider of center-of-mass energy of 100 GeV (or higher), i.e. the kind of accelerator proposed by Alikhanian and Orlov.

At that time there was a complete taboo on Orlov; though he was still free, his arrest was less than a year away. Being aware that a paper containing such arguments might not be included in the program, I gave it a vague title: "Physical processes at energies of the order of 100 GeV in the center of mass". The meeting would be chaired by Logunov, the program was being compiled by S.S. Gershtein, so it was to him that I gave the title of my paper. No objections were raised and my paper was included into the program.

On arrival in Protvino, I told Gershtein what exactly I intended to say. He was terrified: "No, no, do think! Anatoly Alexeevich[18] will be very displeased! It won't be difficult for you to talk about something different!" But I refused. Since my paper was already included into the program, it was no longer possible to cancel it. But Logunov turned out to be more intelligent than Gershtein thought. After my talk he said: "It is good to have differing opinions expressed at our meeting". All other participants expressed support for the IHEP project. A couple of years later CERN initiated its LEP project, LEP was launched in 1989 and outstanding results were achieved on it. The 2 TeV proton accelerator never got built, but this is another story.

An interesting detail. In my talk at the IHEP Science Coordinating Committee session (these talks were published as IHEP preprints) I proposed a mechanism of a search for the Higgs boson H using an e^+e^- collider — the associated production of H and a Z^0 boson: $e^+e^- \to Z^0 + H$, and made an estimate of its cross section. This process is favored by its large effective coupling constant $\lambda\, m_W/e$, giving a large cross section. I told Bjorken about my proposal when he was in Moscow in summer 1976, and he, in turn, reported on it, with reference to my work, at the 1976 Summer School of the Stanford Linear Accelerator Center (SLAC, USA). V.A. Khoze and I wrote a review of possible experiments on a e^+e^- collider at an energy of $\sim 100\,\mathrm{GeV}$, in which we discussed the associated ZH production. This article was first issued as a preprint of the Leningrad Institute of Nuclear Physics (LNPI), and then submitted to the journal *Physics of Particles and Nuclei*. The paper was held up for nearly two years and appeared only in 1978. The associated $H + Z$ production became the principal method of the search for the Higgs boson at LEP. Our LNPI preprint was known by hardly anyone (let alone my Protvino talk), but Bjorken's lecture was read by everybody. Therefore the process $e^+e^- \to ZH$ became known as the Bjorken process, although Bjorken repeatedly said that he was not the author of this proposal. But the reference to our paper with Khoze appeared in the *Review of Particle Physics* of the Particle

---

[18] A.A. Logunov.

Physics Group only in the 2002 issue.

Together with L.A. Artsimovich, Alikhanian was one of the first people in the USSR to engage with problems concerning the exhaustion of the Earth's resources and environmental pollution. He discussed these problems with colleagues, he provided relevant materials, he made attempts to bring these problems to the attention of those in power.

I should like now to discuss Alikhanian's activity at the USSR Academy of Sciences. He was a Corresponding Member of it, and in this capacity he played an active role in creating the Department of Nuclear Physics of the Academy (DNP AN). (It is well known that this was created at the initiative of Academicians Alikhanov and Veksler.) At elections of new members of the Department, Alikhanian showed great persistence in making sure that only physicists of the highest caliber were accepted and refused all compromise.

In the Academy of Sciences, Corresponding Members count as second-class citizens in comparison with full members of the Academy — Academicians. This was the situation in former times, this is also the situation now. At the time when Alikhanian was already a Corresponding Member, the Academy followed a rule that had been established at some early date: both Academicians and Corresponding Members were elected only by Academicians. Alikhanian and Artsimovich established "the ecological principle" during discussions about who had the right to elect Corresponding Members: they declared "Any animal in the world has the right and the opportunity to reproduce his equals". It was difficult to contest this, and Corresponding Members were granted the right to elect their equals.

But in other matters the inequality between full and Corresponding Members remained. Alikhanian was subjected to it himself and very much wished to be elected to full membership. His achievements in science were without any doubt worthy of this. Voting sessions succeeded each other, but still he was not elected. He felt this very strongly, so strongly that it affected his health — alas, he was not free of human weaknesses.

Alikhanian hated the fact that members of the Academy of Sciences were subject to supervision and petty management by State and Party bureaucrats. These sometimes were grotesque. He used to tell the following story: He once spent some time in the sanatorium called "Uzkoie" which belonged to the Academy of Sciences. At that moment there were mainly elderly or ailing Academy members at the sanatorium, often accompanied by their wives. In the evenings, those who were capable of it would come to the common sitting room to sit there and chat. But the sanatorium had (routinely) employed a specialist in mass entertainment, a sort of cheerleader, who would come in the evening and force those present to learn a song:

> We need Academicians —
> yeah, yeah, yeah
> Academicians defend our country
> — yeah, yeah, yeah!

All those present had to sing the refrain "yeah, yeah, yeah" — "They were not allowed to do anything else", said Alikhanian.

Just like his brother, Alikhanian did not like the Soviet regime. The roots of this aversion were deep. In the 1930s he lived in Leningrad, sharing a room with L.A. Artsimovich — who remained his close friend throughout their life. In 1937, the two friends always tried to return home as late as possible, preferably in the morning — Alikhanov used to say that this was aimed at tricking fate. At the same time, however, Alikhanian was aware of the nature of the world in which we all had to live, thus he became a skilled politician, he knew how to navigate through political currents, while avoiding (most of the time, anyway) to act against his principles.

Alikhanian had one weakness: he was susceptible to flattery, and some staff at YerPhI used this weakness to advance their careers. Later, when his position as Director weakened, they abandoned him and joined his opponents.

One last detail: it was Alikhanian himself who chose for the Moscow ITEP the building which it still occupies (a large house with columns) and he himself also found the man (from Ossetia, who was at the time in detention) who recreated the ancient stucco decorations in that building (and was rewarded for it by an early release from prison).

3.5. A.B. Migdal

"Migdal may be late, but Migdal never lets you down"

A.B. Migdal

I heard these words many times from Arkadi Benediktovich (Migdal), or A.B. as many people called him. And he was right.

A.B. was one of my examiners in my doctoral thesis. The situation was not an easy one. The thesis consisted of two parts. The first part was on weak interactions. There I had proved that if parity (i.e. left-right symmetry) was violated, then necessarily charge conjugation or time reversal symmetry must also be violated, and the odd-P pair correlation of spin and momentum of a particle are possible only if the C-parity is also violated. (This result was found before the Wu experiment in which parity violation was discovered); I had also shown the connection between $\pi^0 \to 2\gamma$ and $\pi^- \to e^-\nu\gamma$ decays and much more. The second part was on strong interactions, and in particular on dispersion relations, and there was a danger lurking in this. I had derived the dispersion relations by my own method, based on physical considerations — the Huygens principle and of course causality. (The idea of this method had been suggested to me by Landau. Therefore there were two authors of the paper in the original version; then Landau withdrew his name, saying that he had not done enough on this work and therefore could not be an author.) In addition to the well-known dispersion relation for pion-nucleon scattering, I had found also a new one for nucleon-nucleon scattering. (This was at the same time done by V. Feinberg and I. Fradkin; Pomeranchuk, in his famous theorem of the equality of the cross-sections of particle and antiparticle scattering, referred to their paper as well as mine.) But by working on the derivation of the dispersion relations, I was trespassing on alien territory. It was thought that the only correct method of deriving dispersion relations was the method of N.N. Bogoliubov and his school. And although I emphasised that my method was not rigorous but heuristic and that its advantage was that this method enabled one to obtain results which could not yet be found in Bogoliubov's method, there was a serious danger that my thesis could be sunk by using a "black opponent". (By the way, by Bogoliubov's method only two dispersion relations have been proved rigorously — those for πN and for $\pi\pi$ scattering.) Therefore Pomeranchuk, on whose initiative I had started writing up my thesis, said: "We need strong opponents!" (At that time — I was writing the thesis in 1960 — nobody at ITEP started writing his thesis on his own initiative but only after the go-ahead by Pomeranchuk, and not just after the first, but after the second or third prompting.) Chuk immediately also named the opponents: Migdal, Zel'dovich, Markov. They were all Corresponding Members, and in those times that was a very strong group. Chuk had a word with them and they accepted. Here I must say that Markov let me down eventually: two days before the defense, he sent his conclusions, but only on the first part of the thesis. His verdict was positive and it said that the first part was sufficient for the award of the title of Doctor of Science. He also said that he was going away on holiday and would not be present at the defense. By the rules of those times, the presence of all three opponents at the defense and their personal statements were compulsory — if even one of them was absent, the defense was cancelled. The situation was further complicated in that by the same rules the defense could not take place at the Institute where the candidate was working: the defense had to be conducted in a different institute and one of the opponents had to be from that institute. My defense had to be conducted at FIAN where Markov was working. Feinberg rescued me from this impasse: he agreed to be an opponent and in a single day wrote his conclusions, within one day was appointed as opponent, right before the defense.

But let us return to Migdal. At our very first meeting he told me that he was very happy to be my opponent. For a long time he had wanted to learn quantum field theory and anything new in particle physics, and he hoped that he could learn this by studying my thesis. I replied that I was ready to share all I knew with him. "You and I will meet many times, but there still is time", he added. "Of course, we have at least six months before the defense", I answered. In fact, it all took more than a year, since at that time new rules were introduced, the defense was moved to FIAN, etc. Whenever I met A.B. he told me that he was just about to start reading my thesis, he would then call me and we would do a great deal of work together, but there was still time, wasn't there? At last, with only two weeks remaining before the defense, I telephoned A.B. myself and asked whether I might be of use to him. "Yes, yes, of course," said A.B, "do phone me early next week". I rang. "We absolutely must meet. What about this Thursday? But do phone first". I rang on Thursday. A.B. was absent all day, he only turned up late in the evening. "Let us meet on Saturday, do phone me around 11 am". I rang on Saturday. A.B suggested that we meet on Sunday at 12 o'clock. I phoned on Sunday at 11. His wife told me: "Arkadi Benediktovich has gone to

the swimming pool, do phone in the afternoon, around 3–4 o'clock".
I phoned in the afternoon. His wife said: "Arkadi Benediktovich is
asleep. Do phone around eight". I get through to him at last at eight
o'clock. A.B. invited me to come at nine. I arrived. A.B. greeted
me with delight and explained: "I was aware of a great and difficult
task ahead which meant that I had to be in good form. So I decided
to go to the pool in the morning and to have a good swim. Once
I came back I sat down to lunch and I felt like having some vodka.
After the vodka I felt like sleep. But now — now, you and I will have
a good working session!"

On the following day, Migdal arrived at the Learned Council on
time, and he did bring his conclusions. Migdal had not let me down!

To end this story — though this no longer is about Migdal: here is
the trick by which the danger of a "black opponent" was circumvented. One had to choose a "neutral"organisation and that was to be
the JINR Laboratory of Theoretical Physics which was at the time
headed by Logunov. (Logunov belonged to the Bogoliubov school.)
The reasoning was this: either Logunov writes a negative conclusion,
i.e. he decides on an open conflict, or he gives a positive conclusion
and thereby cuts off a possible negative conclusion for a "black opponent" from the same stable. The trick worked — after several
conversations with me, Logunov gave a conclusion which was bitter-
sweet, but still positive. As I later heard, they did appoint a "black
opponent", namely D.D. Ivanenko, but he could no longer do anything apart from retaining my thesis in his keeping for a year and a
half.

Although A.B.'s knowledge of quantum field theory was limited
in 1960, he soon filled this gap. He was the first to introduce the
Green function method into nuclear theory and using this he proved
that for energies lower than the Fermi momentum it is possible to
describe the nucleus as a gas of interacting quasi-particles. Using this
path, A.B. and his pupils obtained several results in nuclear theory.

A.B. has many achievements in science. I would like to write
about those which concern me most. First of all, this is the ela-
boration of the theory of bremsstrahlung in matter — the so-called
Landau-Pomeranchuk-Migdal effect. It is well known that Landau

and Pomeranchuk had noticed that in matter, the longitudinal distances over which bremsstrahlung takes place increase with increasing energy of the radiating particle, at sufficiently high energies exceed the inter-atomic distances, and coherent radiation on many atoms arises (1953). But they considered only soft photon radiation and used only classical theory. (Earlier on, a similar effect in crystals for the emission of very soft — optical — photons had been studied by M. Ter-Mikaelian.) In 1956, Migdal developed the theory of coherent bremsstrahlung in matter for photons of any energy. The problem had become a quantum one, and for its solution he had to write down and solve the kinetic equation for the quantum density matrix, which before him nobody had done. Chuk said about this work: "A.B. has greatly perfected himself!" I, for my part, can also judge how difficult this problem was. In 1952 I had to solve the problem of propagation of γ quanta taking account of their polarisation in a completely ionised gas at high temperatures, comparable to the electron mass. Here one also had to construct and then also solve the kinetic equation for the density matrix of gamma quanta, but this equation was quantised only in the polarisation variables of the photon, i.e. the density matrix was a 2-by-2 matrix, and the spatial dependence was described by classical equations. I can imagine how much more difficult was the problem solved by Migdal, in which the spatial dependence was also given by wave functions.

Another paper by A.B. which I liked very much is the so-called Migdal-Watson effect: taking into account the final state interaction of pion production in pp collisions: $pp \to \pi^+ pn$ or $pp \to \pi^+ D$. That work was done in 1950, when the first data on pion production in accelerators became available. Migdal showed that taking into account the final state interactions at low energies above the production threshold (and all data at that time were low energy data) reduces to the well-known S phase shift of pn scattering, and found a relation between the cross-sections of the above reactions. His formulæ agreed perfectly with experiment.

About a year later, Watson did some similar work in the US. A.B. was at that time working in LIPAN. His work was secret and he could not obtain permission to publish it. A few months later, A.B. came to Kurchatov (the Director of LIPAN) and with the words: "Here is what one is forced to do when one is prevented from publishing one's work on physics", he deposited a book on the Director's table. The cover read:

A.B. Migdal and M.V. Chernomyrdik,
"How to care for and train reptiles".

The book started with the words: "Everyone likes to have in his home a kind, friendly and obedient reptile".

Even this did not help. A.B's article was published only in 1956, so that until 1956 Watson was the only known author of this beautiful effect. (Members of the Landau seminar knew about this work of Migdal's, but one could not mention it ever.)

Migdal came close to discovering the theory of superconductivity; he was predicting that the cause of superconductivity is related to oscillations of the lattice. When the isotope effect in superconductivity was discovered, and immediately after it was announced, A.B. happened to meet Chuk in the street, and Chuk, without saying a word, lifted his hat to him. When A.B. was writing up his memoirs about Pomeranchuk,[19] he wanted to include this episode and asked me: "Won't it appear immodest of me to write about this?" I assured him that it would not, that it was precisely this sort of detail that showed the real Chuk — the man and the scientist. A.B.'s hesitation and his question to me were both essential details in this task.

Migdal took part in the atom project, he made a significant contribution to the theory of nuclear reactors.

Chuk loved A.B, he always spoke warmly about him and had the greatest trust in him. Here is one example of this. At one time, Migdal had a motorcycle phase and even used his motorcycle to travel to and from Dubna. Chuk, on the other hand, never engaged in any sport at all and generally speaking was an exceptionally cautious man. One day, Chuk told me: "I came back yesterday from Dubna on Migdal's motorcycle". I was amazed. Chuk went everywhere with the fattest possible briefcase. Did this mean that he had had this

---

[19] A.B. Migdal, *The Pomeranchuk Legacy*, in *Remembering Pomeranchuk*, Moscow, Nauka, 1988 (in Russian).

briefcase with him on his motorcycle ride? "Yes," said Chuk, "and what is more we were singing at the top of our voices as we entered the town of Dmitrov!"

Now I would like to tell yet another story, which confirms the principle followed by A.B. — the one I mentioned in the title of my reminiscences about him — the story of how A.B. climbed up the Abdukagor pass.

In 1967 we decided to go on a climbing hike in the Pamir mountains, to the upper reaches of the Fedchenko glacier and we decided to reach these via the Abdukagor pass. This is a classical route for reaching the upper reaches of Fedchenko which allows to by-pass a lengthy climb up the glacier itself. "We" meant A.L. Lubimov (Alik), his colleague Emelin (Igor), B.V. Geshkenben (Boris), A.V. Gurevich (Alik-2) and I. Apart from Lubimov none of us had ever been in the Central Pamir and at great altitudes, and here the upper reaches of Fedchenko were at about 5300 metres and the hike would start at 4300 meters. Part of the climb towards the pass up the glacier had to be done in mountaineering boots with crampons, and apart from Alik, all of us had minimal experience of this. Therefore he naturally became the head of our group.

Alik planned our hike most carefully. We actually had three four-man tents for the five of us! It was envisaged to take the pass after 7 to 10 days of training walks, which would include a gradual increase of altitude. At our first climb to the pass we would move our tents there and also some of our equipment, and we would then return to base camp on the same day, where we would have a few days' rest in the two remaining tents, and then we would start on the final climb to the pass, taking another tent with us. There was no firewood at all on our route, so we were carrying cans of petrol and Primus stoves (that worked on petrol). We bought 10 large cans of concentrated lemon juice (to prevent scurvy). Alik advised us to bring white shirts: he maintained that it was good to wear a white shirt while walking on a glacier (which would be covered in snow) because it would reflect the rays of the sun.

Shortly before we left I heard that A.B., accompanied by another physicist and a young woman, were also leaving for the Pamir mountains. A.B. intended to lecture to the frontier guards, and to show their gratitude they would take him round all sorts of interesting places.

We took a plane to Dushanbe. I will not enter into the details of our further progress. We eventually reached the starting point of our hike: the settlement used by geologists, called "Dal'ni" (i.e. The Far One). Altitude 3600 metres. Here the road stopped and cars could not go any further. Actually, the road used to be longer: it continued along the river Abdukagor and led uphill to a place where quartz was mined. A few years before our hike, the Medvezhy glacier moved and cut the road off above Dal'ni. The geologists cut a donkey passage across the glacier, but a car could not pass this place. We put up a tent near Dal'ni (one tent was enough, as Boris slept outside) and decided, in the interests of training and acclimatization, to spend a few days walking on the Geography Society glacier — its tongue was one kilometer away from our camp.

Late one day, when returning from the glacier, we discovered that the entrance to the tent was blocked by a huge boulder. We had great trouble removing it (how on earth did they manage to get it there?!) On top of the boulder sat an old boot and under the boot we found a short poem (4 lines) written by Migdal in imitation of a rubai by Omar Hayam with signature:

> Yes, there is a sad end to the book of life.
> Do let red wine adorn the flitting of the pages...
> You will not find life's reality in science,
> But rather — discovery will come from a mysterious glance.

We went to the geologists' abode — Migdal and his team were staying there. It turned out that A.B. intended to climb up to the Abdukagor pass just as we did. In my considered opinion, an acclimatization to high altitude was essential, therefore I tried to talk him into staying in Dal'ni for a couple of days, since A.B. had never before experienced high altitude. He said however that his frontier guards would come and collect him six days later, so that he and his group would go up with the caravan of a large mountain climbing expedition which was starting off on the following day. As for us, we spent another two days in Dal'ni and then went up to base camp with another mountaineering caravan — fortunately they allowed us to place some of our load on their donkeys.

So we arrived at our base camp. The camp was situated at altitude 4300 meters in a pocket of the Abdukagor glacier's moraine, that was between the moraine and the rocks that surround the glacier. The exit from the camp was upwards onto the glacier covered with a mess of stones. The glacier was divided into two streams by a nunatak,[20] bearing the wonderful name of "Peak of managing materials". The camp was full of people: two or three mountaineering expeditions intending to climb the peaks and glaciers around Fedchenko. This was part of a national Soviet championship competition. There was a group of members of the Mountaineering section of the Moscow Hiking club, and some other people. A.B. greeted us and explained that he intended to spend another two days there, walking about on the glacier, getting used to walks in mountaineering boots with crampons and then he would go to the pass with a group of mountaineers which was to take some stuff up there to be kept for further use. When I said that I had hardly any experience in hiking on crampons, he immediately said "come with me, I shall teach you". Naturally, our whole group also decided it wanted to learn from Migdal.

I was on kitchen duty the following morning and therefore I got up very early. Migdal and his team were still asleep, both tents were closed. I started preparing breakfast so that we would not keep A.B. waiting when he would get up. A.B. soon appeared and we all set off onto the glacier. A.B. taught us how to jump over crevasses, using our ice picks to ensure our safety, he showed us how to climb a steep slope using the front gripping devices. He did all this with a certain grace and even in this we could detect his teaching talents.

Early on the next morning A.B. set off on the climb towards the pass with a group of mountaineers who were taking supplies to a cache there. He returned at dusk, walking with difficulty, supported by one of the mountaineers. A little later I crawled into his tent.

---

[20] A nunatak is an exposed rock on a glacier (also called *glacial island*).

A.B. was lying there in his sleeping bag. "I am not well," he said, "I am shaking all over, I must be running a temperature." I brought a thermometer and took his temperature which was 37.2 (almost normal). I took his pulse. The pulse was fast, but within acceptable limits. A.B. asked me "What do you think, I am not going to die?" "No, of course not," I said. "You are over-heated and over-tired: after all you have been working all day in the Pamir sun, plus with the light reflected from the glacier, and in any case you have not had enough training or acclimatisation. You will feel very much better tomorrow." On the following day A.B. did indeed feel better and he told us about their ascent. Here is his story:

"At first, I walked well and did not fall behind the others. We went through the part of the glacier that is covered with its stone moraine and also the open black glacier.[21] When the white glacier with hidden crevasses began, the mountaineers showed me the safe route at all times. The snow was deep, walking was hard and I got tired. Then we had to climb up a steep slope. And here I felt that I could not walk at all — I would stop right now, I won't go any further. But everybody walked on .... I must walk too. And I

---

[21] The climb to Abdukagor pass from base camp leads first over a part covered with a stone moraine (stones of different sizes, some frozen in, some unstable and moving, on ice-hummocks). Then through the black glacier zone, it is criss-crossed with open crevasses — one has to circumvent or to jump over these. Here one must use crampons, because the slope is at 20 degrees and boots just do not hold fast. After this comes the white glacier zone which has hidden crevasses. The angle of the climb is not steep here so one does not need crampons, the main danger comes from the crevasses. Therefore everybody ropes up and one moves ahead in groups of 3 to 5 people, all tied together. It is very difficult to see a crevasse when it is covered with snow, an ice pick does not necessarily feel it and in any case you would not use your ice pick every 20–30 centimetres. Therefore people very frequently fall into crevasses. (When our group was climbing towards the Abdukagor pass, this happened several times and I fell into a crevasse once or twice myself.) Those who are tied to your rope immediately ensure that you are safe by sticking an ice-pick deep into the ice and winding the rope around it. In fact, this precautionary measure is not needed in practice, one does not fall into the depths because of one's rucksack. When the rucksack touches the snow, the pressure immediately reduces and the snow holds. The experience remains very unpleasant, however, one feels one's legs dangling in a void. One begins to crawl out — forwards or sideways — but it sometimes happens that the snow collapses under you again, and one no longer knows whether one is crawling across the crevasse or along it. After this glacier zone you have a relatively minor slope (20–30 degrees) covered with névé and this leads you to the Abdukagor pass (5100 metres). Immediately after it you see the huge expanse of the Fedchenko glacier, on the horizon it is circled by a wall of 6000 meter peaks.

would take one step, another step .... I could already see the pass. I felt — I can't get there, this step is the last one. I get the very last strength together, I reach the pass and I hear the command: 'Fall in'. I think 'Right, I shall get in line, and I'll fall over and die .... But this is an honourable death', and I take my position in the line-up. The head of the group, a Master of Sport, says 'I congratulate you on reaching the Abdukagor pass'. And I stand there, thinking — 'It's all right to die now!'".

And this is how the head of the group, the Master of Sport in Mountaineering, described this climb on the following day (Migdal had already started on his return walk):

> "To start with, that old guy, the Academician, was not too bad, he almost kept up with our people. Well, our people were of course carrying the equipment, their rucksacks held 30–40 kilograms each. He did go soft after a while, that's true. But my boys — they can drag a cow up to the pass, sure thing. He did get there, yes, he did climb right up. Well, going down, of course, he had to get some help."

That man was not aware of the main message — Migdal had not let anyone down!

Now I want to tell about the events that followed three days later. Although they are not directly connected with Migdal, since by then he had already gone away down, an account of these events that complements A.B's climb rather well. One morning we were woken up by some screaming: it turned out to be a man who had run into the camp and told that a girl in a hiking group on the Fedchenko glacier was dying. She had developed pneumonia, but the head of their group, who had undertaken a hike of the 5th (highest) category of difficulty, did not want to cut this short and the group continued to advance — and so did the sick girl. (Pneumonia is a terrible thing in the mountains at high altitudes: there is little oxygen there, antibiotics do not work, and it takes just days before death strikes. The only way to survive is for the patient to be brought down immediately from high altitude.) The girl's condition got worse and

worse, and in the Abdukagor pass area the head of the group decided
that the whole group would continue its hike, but two people —
together with the girl — would go downhill with her, following the
Fedchenko glacier. The girl could no longer walk and the two men
had to carry her as best they could. The man who had run to the
camp was one of them. All those who could do so immediately got
ready and stomped up the glacier. There were no mountaineers left
in the camp at the time, but we had a group of the mountaineering
section of the Moscow hiking club — these were mountain hikers of
the highest class. They ran ahead of everybody else, carrying ropes.
I got dressed, grabbed a board that was lying nearby and set off
at a run. It really was not easy to run upwards at an altitude of
4500 metres, on a glacier covered with stones, and with a board, too.
When I reached the rescue party who were carrying the girl and saw
the girl, I took fright — I thought I was seeing a dead body. Her
face was yellow like that of a corpse, she did not react to anything,
only a shallow breathing showed that she was still alive. Somebody
had brought another board, so the two together were made into a
sort of stretcher, the girl was tied to this, and the descent along
the glacier went on. It took 10–12 people to carry her, because one
had to maintain a difficult balance among the stones and the ice-
hummocks. Another three or four people were running ahead of the
party, to find and show the best route to follow. That's how we made
it to the camp. Here the girl felt a little better, she opened her eyes
and said a few words. She was given a shot of antibiotics and a hot
drink. It was decided to transport her immediately to Dal'ni, and
two men ran off to Dal'ni in order to use the emergency radio set to
call up a rescue helicopter (there was no radio link in the camp).

The only way down lay over the moraine, along a steep and winding path, which was entirely impassable for our makeshift stretcher.
So this is what was done. A large rucksack was taken, two holes were
cut into it, and the girl was inserted into the rucksack with her legs
thrust through the holes. Then two men lifted the rucksack with
the girl, loaded it onto the back of a third and tied the top of the
rucksack to his neck. The girl turned out to be quite heavy — about
65 kilos. It was not possible to carry such a weight for more than 10

minutes, after that she would be moved to another porter. Ten or 12 people, including Boris and myself, set off downhill with her. It was interesting to see how life was coming back to her as the altitude decreased. When she was loaded onto my back for the first time (this was not so far from the camp), she was almost lifeless. But when my turn came round again — this was at about 3800 meters, on the disused road — she could already talk to me and even make jokes. Soon after, we saw the people whom we had sent to Dal'ni. They said that the rescue helicopter was coming. The girl was safe.

This story shows that A.B. did have serious reasons to worry after his return from the pass — I had not realised this at the time.

3.6. V.N. Gribov

No man is a prophet in his own land

Vladimir Naumovich Gribov was without any doubt the greatest theoretical physicist of the post-war generation in the USSR. Even a simple list of his main achievements in science is impressive:

> the theory of many particle reactions near threshold;
> the Gribov-Foissard representation;
> the shrinking of the diffraction cone at high energies;
> the factorization of the Gribov-Pomeranchuk Regge pole contributions;
> Gribov-Morrison selection rules;
> the Glauber-Gribov theory of diffraction scattering on nuclei;
> the Gribov-reggeon diagram technique;
> the Abramovski-Gribov-Cancelli rule;
> the Bjorken-Gribov paradox and the Gribov generalised vector dominance;
> the Gribov-Pontecorvo neutrino oscillations;
> the high energy bremsstrahlung theorem;
> the Gribov-Lipatov structure function evolution equations;
> Gribov copies and much, much more.

Gribov did not participate in the atom project but he established

V.N. Gribov

a creative atmosphere which spread well beyond the areas in which he did his work.

In particle physics, he did more than anyone else in our country. But at all times and in all lands "no man is a prophet in his own land", and during Gribov's lifetime his achievements did not receive the recognition that they deserved in the USSR (and they did not get enough recognition abroad either). He was elected Corresponding Member of the Academy of Sciences of the USSR very late indeed, much later than quite a few other theoretical physicists, and the Academy did not find a place for him in the ranks of its full members right until the end of his life.[22]

He was awarded only one of all the many prizes and other marks of distinguished achievements which existed in the USSR and later in Russia: the Landau medal. It is true (as he told me himself) that this was the only award that he did wish to receive. Also it was very rare for Gribov to be asked to present one of the prestigious summary reports at major international conferences (and only once in the USSR — at the Dubna conference in 1964).

There were however men who highly valued his talent straight away — Pomeranchuk and Landau. In the late 1950s, N.N. Bogoliubov's name was put forward for the Lenin prize, based on work on dispersion relations. Landau was sent the materials on which this proposal was based. Landau wrote in his conclusion that the Lenin prize for work on that particular theme ought to be awarded to Gribov rather than to Bogoliubov. Shortly before this, Gribov had done work on the spectral representation of vertex functions in field theory, and Landau considered this to be a much greater achievement than Bogoliubov's proof of dispersion relations. I must note that this happened at the very dawn of Gribov's creative work, much before his famous work on Regge theory and everything else! Of course, Landau's conclusion had no effect on the Lenin Prize award Committee and the prize was given to Bogoliubov. It is easy to comprehend the consequences Landau's conclusion had on Gribov. He was aware of

---

[22]Gribov had been elected Corresponding Member of the Academy of Sciences in 1972 and he died in 1997. During that time eight theorists were elected to full membership in the theoretical Physics sector to which he belonged, but Gribov was not considered worthy of that honour.

them for a long time, even possibly to the end of his life.

As to Pomeranchuk, he not only valued Gribov, he also was very fond of him. I remember the heroic times at the start of the Regge theory work, the work that Gribov and Pomaranchuk did together (they had 14 joint publications). Whenever Gribov came to us in Moscow from Leningrad, this was a real cause for celebration for Pomeranchuk — well, for all of us at ITEP. Discussions would start in the morning and lasted well into the night. Pomeranchuk's small office would be blue with smoke: both he and Gribov never stopped smoking. And after a few days' work Truth emerged from Chaos — a wonderful delight!

For Pomeranchuk, Gribov's opinion was extremely important, almost to the same extent as Landau's opinion. A characteristic example: our work with Gribov and Pomeranchuk on the behaviour of the e^+e^- annihilation cross section into hadrons at high energies — Pomeranchuk's last work. This work is special in his oeuvre. After the proof of the null charge in quantum electrodynamics and in meson theories, Pomeranchuk (as well as Landau) considered that "The Lagrangian is dead and must be buried with full honours" (Landau's words). For 10 years, Pomeranchuk developed phenomenological methods and methods based on analyticity in particle physics (Pomeranchuk theorem, Regge theory, SU(3) symmetry and others). In the work mentioned above, Pomeranchuk revisited the methods of quantum field theory, i.e. the Lagrangian. Such a revisit was difficult for him, because he wanted to be sure that Gribov completely shared his point of view. Pomeranchuk was already seriously ill (cancer of the digestive tract) — he could not swallow, he spoke through an amplifier, but he worked, wrote formulae! We discussed with him, at times Gribov came from Leningrad. In one of our discussions — this was about two weeks before Pomeranchuk died — he and I (Gribov was away from Moscow) came to the conclusion that the job was done, we had the result. "But" said Pomeranchuk "do telephone Gribov and if he agrees with everything, do start writing up our article". I phoned Gribov and he said that he had developed some doubts concerning the proof. I told Pomeranchuk. His reaction was — although he had no doubts himself — that it was not possible to proceed as long as Gribov had even a shadow of a doubt. I telephoned Gribov again and asked him to come to Moscow immediately. Once he had come, he and I discussed our proof and spent a couple of days on these discussions. At last we found something which removed all possible doubt. After this we went to see Pomeranchuk. It was on Sunday the 12th of December, 1966. "Volodia" asked Pomeranchuk "have you got any doubts left?" "No" answered Volodia and it seemed to me that a shadow of relief flitted across Pomeranchuk's face. But the conversation was brief, Pomeranchuk did not feel well. He died on the night of Tuesday, 14th December 1966. The article had to be written without him.

During discussions on science with Gribov (and in fact not only on science) one always felt a high degree of creativity, he was a burning light (I cannot find anything better than these commonplace words). At the same time, he was totally intolerant of compromise in science. It was utterly impossible to force him to agree with any scientific work which he considered incorrect, or even to keep silent about it. But if as a result of discussions (very lengthy ones at times) Volodia gave his agreement, one could be 100% sure that the research was entirely correct. This obviously had its reverse side. At times, Volodia's judgement would be mistaken and he would not accept an idea which was in fact correct and occasionally very good. Since his arguments would be persuasive (though later shown to be wrong) and since his prestige and authority were high, a man would lose courage.

Here is one example that is very important to me and one that saddens me: in the beginning of 1972, after 't Hooft had proved the renormalizability of nonabelian gauge theories, I understood that the arguments of Landau and Pomeranchuk on the internal inconsistency of the Yukawa' theories (the unphysical pole of the effective charge at high energies) did not apply to nonabelian theories. The logic of my thoughts was the following: the arguments of Landau and Pomeranchuk were actually based on the Källén-Lehmann representation for the photon propagator in quantum electrodynamics (also in meson theories). According to this representation, since the imaginary part of the propagator is positive, it has to grow with increasing energy, and then the appearance of a pole is inevitable. But in nonabelian gauge theories the propagator of the gauge boson is not gauge invariant, and then one cannot make a corresponding statement. However, I was not in command of the techniques of calculations in nonabelian theories. Here Vainshtein arrived from Novosibirsk, and he was familiar with this technique. I tried to persuade him to carry out the corresponding calculations, I tried for two days, and on the third day I succeeded. But as luck would have it — Gribov arrived from Leningrad and in the course of a couple of hours he made Vainshtein change his mind. Gribov stated with great conviction that there would be the same pole in nonabelian theories (i.e. the zero of the physical charge) as in quantum electrodynamics. To my shame, I have to confess that I had missed an earlier paper by Khriplovich where the calculations that I needed had been done. Vainshtein surprisingly did not mention it to me. Vainshtein went back home. The study of techniques of calculations in nonabelian theories required time which I did not have: soon I had to go to Czechoslovakia for the start-up of a nuclear power station.

Gribov was able to approach a problem, a phenomenon, from a new, unexpected angle, as a rule a deeply physical one. The phenomenon then started to shine in new colours. One can give many examples of this: instantons (the idea that instantons in Minkowski space describe transitions between vacua with different topological numbers belongs to Gribov), Gribov copies and others. Another example, one closer to me: sum rules for γn and en scattering (papers by Gribov, Shekhter and me). Here Gribov could approach this problem from the point of view of the Yang-Mills theory, which greatly helped to understand it. Another similar example: the work by Gribov on photon-nucleon interactions and the connection of deep inelastic scattering with e^+e^- annihilation — the Gribov-Bjorken paradox. To find and to formulate a paradox — and Gribov knew how to do that — is the best way to move science forward.

When Gribov spoke at a seminar he would always speak without notes, he presented his thoughts as if inviting those present to solve the problem under discussion together with him. In this he and Pomeranchuk were much alike: Pomeranchuk also seemed to improvise when he gave a lecture or when he spoke at a seminar. (It was all very different when Landau spoke: there it was evident that as far as he was concerned the problem had been solved and he was explaining it to us, poor ignaramuses.) When Gribov spoke, the seminars both at ITEP and as far as I know also at the theory department of LNPI would almost invariably last well into the night. Now that I mentioned LNPI, I feel obliged to point out that in actual fact the theory department of LNPI was created by Gribov. It is true that its foundations were laid by I.M. Shmushkevich, and these were foundations of a good quality, but the whole building was erected by Gribov and his traditions are still alive in LNPI (now called PNPI). No major work on theory — not only on particle physics, but also on other fields of theoretical physics — could leave LNPI without being discussed with Gribov, and these discussions were always highly fruitful for the authors. He also exerted a strong influence on the experimental research in LNPI.

The situation changed when Gribov transferred to Moscow. I feel that at least for the first few years after he moved this was a hard, possibly even the most dramatic period of his life. Life in Moscow was entirely different from life in Leningrad: here, a major part was played by various relationships which had only little to do with science, sometimes even by intrigues, by hierarchies in science. Some things were not allowed today, others — tomorrow. Gribov did not want to participate in all this, but on the other hand if one ignored all that, life became impossible. His links with the school that he had created in Leningrad were weakening, however much both sides tried to keep them alive. In addition, though science contacts in Moscow did begin to appear, they never were as warm as those in Leningrad. And finally — in Leningrad, Gribov belonged to the overall intellectual elite, not only to the elite in physics, not only in science: he knew many people and many knew him. This did not happen in Moscow. Here the very concept of an intellectual elite is much less defined — much depends on how close a person is at a given time to those in power.

All this was added to the disaster of his son Lionia's tragic, absurd death in the Pamir mountains. Lionia fell into a crevasse on a quiescent glacier and when he was pulled out, he was already dead. I feel a measure of guilt in this tragedy. For several dozen years I used to go trekking in the mountains, then my son followed suit. We were friends with Lionia. It is possible that our example somehow influenced him and made him start this same activity, although he was physically less well prepared for it.

And here I would like to return to the place from which I started. All of us, close friends and colleagues of Volodia Gribov, must feel guilty for the fact that his achievements had not been recognised and valued in Russia to the extent that they deserved. This lack of recognition did of course influence his morale. Applying this to myself, I remember Tvardovski's poem:

> I know, I am not answerable for
> The fate of those who perished in the war,
> For every younger pal or senior vet
> Left yonder; nor is anybody saying
> That I could save them, but fell short of saving —
> No one says that, and yet, and yet, and yet ...

And I would like the words that I wrote here to be taken as my belated expression of repentance.

3.7. Ya.B. Zel'dovich

A flair for theories

Among all the qualities of Yakov Borisovich Zel'dovich the theoretical physicist, there is one which in my view is particularly remarkable and which singles him out among other theorists — this is his flair for theories. I mean his remarkable gift of feeling the depth and perspective of a theoretical thought or idea, when this thought or idea is not in the least formed yet, when it is raw, when it may even look absurd rather than rational, and when everybody else simply ignores it.

I shall give several examples of such foreknowledge.

In 1959–61 Salam, Ward and Glashow began to publish the

Ya.B. Zel'dovich

first attempts at uniting weak and electromagnetic interactions. At that time these articles did not attract general attention, but Ya.B.

immediately noticed them. He would come to see us in ITEP,[23] and he would say: "What a remarkable theory, why are you not working on it?" We would answer that this theory was not renormalizable, it would contain a great likelihood of $\mu \to e\gamma$ decay and so on, but Ya.B. would not desist: in his opinion this theory was so profound that one should work with it anyway, in spite of difficulties. And he was basically right. Although at the time — in 1961–62 — we did not follow his advice, it seems to me that at least as far as I am personally concerned his words did not go in vain: a little later, starting in 1963, I did start to work on a theory with intermediate W bosons.

Another example concerns the 1961 publication by Goldstone, where it was shown that spontaneous violation of symmetry produces massless particles — Goldstone bosons. Everybody at ITEP had the same attitude towards this, namely: everybody agreed that the work was interesting, but no one wanted to develop these ideas further. Maybe this was because almost everybody at ITEP (and particularly Pomeranchuk) was at the time enamoured with the Regge theory. During discussions, Ya.B. repeatedly stressed the depth and promise of Goldstone's ideas and appealed to us to develop them. But, alas, his efforts here were unsuccessful — we continued to follow our own pursuits. But now it is a well-known fact that elementary particle physics is affected throughout by ideas concerning the spontaneous violation of symmetry and the formation of Goldstone bosons.

A third similar example concerns the cosmological term in the theory of gravitation. In the 1970s, Ya.B. began to say that in all existing field theories (including also in models of the Grand Unification) the cosmological term, calculated in perturbation theory, is usually divergent, and even if by using some theory one succeeds in making it converge, its scale turns out to be by several orders

---

[23] In the 50s Ya.B. spent several years working part-time in the ITEP Theory Laboratory as a secondary job, and later he was dismissed on orders from above. Pomeranchuk (who was in charge of the Theoretical Laboratory) and Alikhanov (the Director of ITEP) tried to oppose this decision, but those in power were immoveable: all multiple job holdings got forbidden. The same fate struck Landau — he too was dismissed from ITEP. Discussions with Ya.B. were very useful for us. I still keep as a precious relic a print-out of his 1954 article with a handwritten dedication: "To dear Boris Lazarevich from a grateful pupil". I would like to think that this was not just a joke.

greater than experimental limits. Ya.B. was of the opinion that the choice of suitable theories ought to be decided by the condition that this theory leads to a zero-value of the cosmological term. Ya.B. returned to this idea many times, for him this was equivalent to the call "Carthage must be destroyed!" The problem of the cosmological term remains unsolved, and today the criterion advanced by Ya.B. is one of the principal criteria in the selection of theories of unification of all interactions, including gravitation. There are however some attempts to solve this problem differently within the framework of cosmology. The existence of the cosmological term (or in a wider sense — of dark energy) is no longer subject to doubt.

Ya.B. wrote many papers with co-authors, but in all these papers (I do not know any exceptions) the main physical idea always originated from him. And in accordance with the requirement put forward by Landau (see Section 3.1.5) as a rule, he predicted the result.

In his judgements, Ya.B. was categorical, even harsh. That was natural for a person whose main task over many years had been the creation of atomic and thermonuclear bombs (although this is not necessarily so — Sakharov was a mild man). At the same time, however, one could not call him stubborn: if the reasoning of his opponent was convincing, then he could change his opinion. I remember the following case: One early morning (it was in the 1980s) Ya.B. phoned me (he always phoned at 8 am in the morning, assuming that if he got up at 6 am, so should others). Ya.B. started to try to talk me into dropping what I had been doing because that was not interesting, and rather take up astrophysics. I retorted: "But you don't know what I am doing!" In the next 10 to 15 minutes I described to Ya.B. what exactly I was working on at that time. He changed his opinion. The result was prompt: Ya.B. phoned Khariton and persuaded him into supporting me at the elections for Corresponding Member of the USSR Academy of Sciences.

Another case, typical for Ya.B. Soon after the Chernobyl disaster he phoned me and asked if I would agree to give my comments on the disaster to Yu.B. Khariton. I agreed. Literally after a few hours,

Khariton phoned me and invited me to Barvikha[24] where he was taking a holiday at the time. He took a notebook, made notes, asked questions. I remind the reader that Khariton (together with Zel'dovich) was the author of the very first paper on the theory of nuclear reactors. But as he explained to me, he was not familiar with the contemporary state of that theory. (Khariton at that time was the scientific head of Arzamas-16, i.e. the head of the program of developing atomic and thermonuclear bombs.)

The circumstances of Zel'dovich's life left him very little time for engaging in elementary particle physics, and if his fate had taken another turn he would have achieved much more and also he would have felt much more satisfaction.

Ya.B. Zel'dovich did much work in the Soviet atom project, but his main work in this field remains secret. This work brought him three awards of "Hero of Socialist Labour".

3.8. I.V. Kurchatov

A great organiser and a great scientist in one person

Kurchatov was a very unusual man: an organiser of the highest caliber, I do not know anyone else with such a brilliant gift for organisation. First and foremost, he was influential to a truly colossal degree, though he did not have an official post to justify such an influence. He was the Director of Laboratory 2, which in 1949 was renamed Laboratory of Measuring Instruments (LIPAN) and in 1956 Atomic Energy Institute (IAE). Apart from this, he only held the post of chair of the PGU advisory council on scientific and technical matters at the Ministry of Intermediate Engineering, which had an advisory function. I do not know how Kurchatov managed to become so very influential and I do not know by what means he retained this influence, but there is no doubt whatsoever about this: his influence was retained under all potentates, both under Stalin and under Khrushchev. I shall quote one fact which I witnessed myself. I was present in Kurchatov's office when he needed to telephone Kosygin

---

[24] A sanatorium outside Moscow.

I.V. Kurchatov

on some business. At that time Kosygin had not yet risen to Chairman of the Council of Ministers, but he already was an important figure in government. Kurchatov took the special phone set which was dedicated exclusively to the government communication network, dialled Kosygin's number and said: "Aleksei Nikolaievich, this is Kurchatov. We need this and that to be done. And we need it to be completed by such and such a date. I ask you to take measures for this to be done". And I understood the answer from the other party as: "This will be done, Igor Vassilievich". But at the same time, this man understood and loved science (and I do mean science rather than his own position in science, as is the case of many of our current "organisers" of science).

Here is an episode to illustrate this: it happened in 1955, when the construction of atom power stations and their economic expediency were being discussed. The result of this discussion depended on knowing how much uranium would be needed: how often fresh uranium would be needed to add to the power station, i.e. one had to know the permissible level of burn-up of uranium in the reactor of an atom power station. I was doing the necessary calculations. The complexity of the problem was this: the result strongly depended on physical constants, namely on the parameters of uranium and plutonium, and these were not sufficiently well known. Therefore I took an indirect route and determined the necessary combination of constants, using the data obtained by monitoring the work of reactors producing weapon grade plutonium. I communicated the result of my calculations to Alikhanov, who forwarded them to Kurchatov. There existed another calculation on the burn-up of uranium in nuclear reactors: this was done at LIPAN by S.M. Feinberg. Out of the blue Alikhanov's secretary contacted me (Alikhanov was absent at the time) and said that Kurchatov was on the governmental telephone line, asking for me to speak to him. (At that time I was only a PhD and occupied a minor position at work, so the distance between us was immense.) Kurchatov told me that he was familiar with my calculations and asked me to explain my results. I briefly explained and he said that they strongly differed from those obtained by Feinberg, mine were much more pessimistic and therefore he needed details. I fetched the notebook containing my secret notes and dictated my figures to Kurchatov, who was obviously entering them on graph paper and comparing them with Feinberg's data. The main difference between my calculation and that of Feinberg was that the deep burn-out of uranium (which takes place in atom power stations but not in military reactors) produced an accumulation of Plutonium-240, which was characterised by strong resonance capture. Feinberg did not take this into account (or not enough) because it had never been subjected to dedicated measurements, while I had determined the effective parameters of Plutonium-240 by analysing the working of military reactors. I explained all this to Kurchatov. Our conversation lasted for about 40 minutes, and at the end of it Igor Vassilievich agreed that my calculations were correct, though this was obviously unwelcome for him, since it led to a noticeable worsening of the parameters of nuclear power stations.

Another remarkable trait of Kurchatov's was his amazing ability to find the right people to fit a specific job. An example of this was again the case of Feinberg. He was the head of a group doing physical calculations of reactors in LIPAN. At the same time he was knowledgeable in the field of reactor design and in technical thermodynamics. A combination of these qualities in the same person is extremely important, since the demands of physics and of reactor technology are usually in conflict. Savelii Moiseevich Feinberg had been appointed to his post only thanks to Kurchatov's gift of assessing the value of people at their very first meeting. Kurchatov once had said, while talking to a group of his staff, that he needed a man who could do both the calculations of reactors and deal with the engineering side. One of the people taking part in that discussion was Evgeni Lvovich Feinberg who said that he knew a suitable candidate: a first cousin of his, S.M. Feinberg. His cousin had a degree in construction engineering but he was a very gifted man and he was sure to learn everything about a new profession. He would be able to cope with the task in question. Kurchatov hired S.M. Feinberg on first acquaintance, and his trust proved to be entirely justified.

Kurchatov had this major virtue: though he was the head of the atom program and had the most colossal power, he did not turn into a full monopolist and he did not attempt to crush his competitors, as a contemporary science boss would have done. This can be demonstrated by the example of the program of building atom reactors for producing tritium. Being the head of the whole atom project, Kurchatov could easily have taken this program over. But he did not. He asked his own Institute to submit the project of a reactor working with graphite, and he asked the competing organisation — TTL — to present the project of a heavy water reactor for the same purpose. Both projects were then compared and contrasted. As a result, TTL was not completely crushed, some heavy water reactors did get built. And I think that this was a matter of principle for Igor Vassilievich — he decided to allow a certain volume of competition rather than smother it completely. He also understood that the very existence of a competitor would make his Institute produce better work.

Kurchatov, though, was still a man of his time. He was strict, harsh, a leader who ruled. Monopolism in science came directly from him. One might however say that Kurchatov's monopolism was "an enlightened monopolism", tempered by his understanding that competition was essential, colored by his interest and love for science. (It is interesting to note that Kurchatov joined the CPSU as late as August 1948, by that time he had been at the head of the atom project for over five years.) Among the many proofs of his love of science (and not only of the science in which he was himself engaged) was the creation at IAE in 1958 (at a time when genetics was suffering persecution) of a Radiology Department: some research on genetics was carried out there and some of the geneticists found refuge there. These traits of character were fading in the leaders who came after Kurchatov, also their level of competence in science was lower, so that the tendency towards monopolism was retained and in fact it even grew.

I think that if Kurchatov had lived longer, the Chernobyl (RBMK) reactors would not have been built and the Chernobyl disaster would not have happened.

There is a saying "There cannot be a great man without a great event". It is also true that when the great event which produced the great man comes to its end, the great man departs — and as a rule he departs in a physical sense. It seems to me that this happened also with Kurchatov: by 1960, the immense task of creating nuclear weapons was complete, the problem was solved. There was no space left for him, and he departed.

3.9. A.D. Sakharov

The main scientific achievement of A.D. Sakharov was the explanation of the baryon asymmetry in the Universe (the presence of baryons and the absence of antibaryons), based on CP violation [43]. This paper was cited many times in theoretical work on cosmology and lies at the foundation of a number of cosmological models. Another well known paper by Sakharov is the theory of the magnetic thermonuclear reactor where the nuclear particles which control the thermonuclear reaction are trapped by a magnetic field [44].

However, Sakharov's main achievements are connected with the thermonuclear bomb. They are:

1. "layer cake" ("sloyka" in Russian) — the thermonuclear bomb with layers of deuterium and tritium (later on, V.L. Ginzburg proposed replacing the tritium by LiD).
2. Thermonuclear bomb, surrounded by a casing of heavy material in which γ quanta are produced; the γ quanta, falling on the thermonuclear bomb, produce pressure and thereby raise the temperature.

Sakharov had the idea of placing a chain of powerful thermonuclear bombs in the ocean along the coasts of the USA and of exploding them simultaneously in order to produce a tsunami and the production of an immense quantity of steam and activated water. This would destroy all that was alive at a distance inland of 200–300 kilometers along the coast. This would cause the death of scores of millions, possibly hundreds of millions of people. As A.P. Alexandrov said, he was completely shattered by this statement, which meant the total destruction of people. A.P. Alexandrov called this proposition of Sakharov's "a cannibal idea". Sakharov had put this idea to Admiral Fomin who reacted with the words: "We seamen do not wage war on the civil population". Later on, Sakharov changed his position entirely — he became an enemy not only of the direct use

A.D. Sakharov

of nuclear weapons as such, but also of its long-term consequences.

A.D. Sakharov's work on the nuclear project brought him three awards of "Hero of Socialist Labour".

Sakharov is known to the whole world as a fearless defender of human rights, a promoter of democracy in the USSR. His activity started in 1968 with the article "Thoughts on Progress, Peaceful Coexistence and Intellectual Freedom". The main idea of the article is that mankind has arrived at a critical moment in its history. It is threatened by thermonuclear destruction, ecological suicide through selfi-inflicted poisoning, famine and run-away demographic explosion, dehumanization and dogmatic mythologization (I quote from [5]). This was said in 1968, but remains true to this day. Very soon after this, Sakharov became the leader of the movement aimed at the democratization of Soviet society. He was persecuted by the authorities (surveillance by KGB agents, exile to Gorky), but he became known all over the world. In 1975 he was awarded the Nobel Peace Prize.

I met Andrei Sakharov only rarely as we were working at different Institutes. I repeatedly invited him to take part in the Theoretical seminar at ITEP and in ITEP conferences. He presented at one of these seminars the most famous of his papers, on the presence of baryons and the absence of anti-baryons in the Universe. The paper was followed by intensive discussions. As a result, Sakharov thanked me in his publication.

I fully shared his political views. One episode comes to mind. At the end of a conference day at the Institute of Physical Problems, we came out together. Sakharov asked me:

"Where are you going?"

I answered "to the Centre of town."

"So do I, I intend to take a taxi. Would you like to come with me?"

"Of course."

Actually, Sakharov, a full member of the Academy of Sciences, was entitled to the provision of a car by the Academy, but he rarely made use of it.

In the car, Sakharov asked me:

"Are you acquainted with any members of the Supreme Soviet of the

Russian Federal Republic?"

"No, I do not. But you do know one — Khariton."

"Khariton is a member of the Supreme Soviet of the Soviet Union."

"What do you want with him?"

"I want a member of the Soviet to propose the rejection of the proposed Russian Federation law against human rights activists."

I thought: this is a brave man. He obviously must be aware that many taxi drivers are KGB agents. Moreover, the car may be equipped with a listening device.

I also used to meet Sakharov at parents' evenings in the school where our children were pupils. Entirely different problems were discussed on these occasions.

3.10. B.L. Vannikov

Vannikov was born on 7th September 1897 in Baku, in a working-class family. His father was employed in the oil industry. As a teenager, having completed his primary school education, he was employed as a metal worker, first in the oil industry and later in road building. In 1918 he graduated from the Baku Technological college, in 1926 from the Moscow Higher Technical college (Bauman Institute). He was short of stature, very agile, with a very typical Jewish physique. In his behavior he was sometimes rough and cynical, sometimes very brusque, but also benevolent when needed. In 1927 he was appointed technical manager of a factory, in 1933–36 he was Director of factories — first in Tula, then in Perm. From December 1937 he headed the department in charge of tanks at the Commissariat for Defense, in December 1937 he became Assistant People's Commissar of the Defense industry. From January 1939 he was People's Commissar of Armaments of the USSR.

On 7th June 1941 he was arrested and detained in the Lubyanka prison in Moscow, sharing his cell with Meretskov who later became a marshal, and they both were subjected to merciless beatings. Early in July, while still in prison, Vannikov was instructed to put forward a plan for the evacuation of defense establishments from western parts of the USSR to the Urals and beyond. Vannikov did not know that

B.L. Vannikov

War had started or that the Germans had occupied a considerable part of the country's European territory. Just like the whole population of the USSR, he was expecting us to beat the enemy on foreign territory and with little bloodshed. Nonetheless, his proposals were written down and presented to Stalin. Stalin approved the proposed actions and ordered Vannikov to be released from custody. Vannikov was handed a certificate from the State Defense Committee (GKO) No 1021, which stated: "GKO certifies that Vannikov Boris Lvovich has been temporarily subjected to arrest by the organs of NKVD, it has now been established that this was due to a misunderstanding and that Vannikov Boris Lvovich is to be considered fully rehabilitated. By decree of the Central Commissariat of the Communist Party and the Council of People's Commissars of the USSR, Vannikov is appointed to the post of Assistant People's Commissar for armaments and by order of GKO he must immediately set to work in the post of Assistant to the People's Commissar for Armaments (Ustinov D.F.) Signed: Chairman of GKO I. Stalin".

Vannikov carried out an immense and impressively efficient operation in this post. All the factories to be evacuated were transferrred to their new locations on time and started working there immediately, "off the wheels". It is thanks to Vannikov that the Red Army was provided with the weapons it needed, although the territories where these factories had been located were under German occupation. On 3 June 1942 Vannikov was rewarded for this operation with the award of "Hero of Socialist Labour". (He was among the first six people to receive this new award immediately after it was created.)

Nonetheless, at the end of 1941 Stalin ordered Vannikov to be arrested again and he was sent to a labour camp in the Far North.[25]

Vannikov was set to logging in the forest. One evening, on returning to his camp hut, he was immediately called to report to the camp commander. No explanation was given. Vannikov's request to allow him to fetch his personal possessions from the hut was refused, which augured trouble, and he left the camp in a very depressed mood. He was flown from the nearest airport to Moscow, but no

---

[25] Here I follow the text of G.A. Sosnin, ISAP, n. 2, p. 189, who heard this story from Vannikov himself.

explanation was given. Only after he was driven past the Lubyanka and approaching the Borovitsky gate of the Kremlin did he realise that he was being taken to Stalin, since no one apart from Stalin could have called Vannikov to the Kremlin. He entered Stalin's office and remained standing by the door. Stalin was walking back and forth along the central table, looking down. After a while he started talking in a low voice about the situation on the fronts, about the difficulties experienced by the armies (particularly about the supply of ammunition). Only at that moment did Vannikov realise that he had not been called to be punished. Stalin stopped and said: "We decided to organise a People's Commissariat for Ammunition and we ask you to head it". To this Vannikov pointed at his prisoner's clothes and felt boots and asked how he could possibly stand at the head of a Commissariat if he was a detainee in prison?

Here Stalin looked at him closely, as if he was seeing his clothes for the first time, limply waved his hand and said that Vannikov ought not to worry about this. He advised him to go to the hotel and to come to the Kremlin within three days with his proposals concerning the organisation of his Commissariat. The fact that Vannikov was not sent home from the Kremlin but rather was to go to a hotel meant that although he had been granted freedom to discuss matters with the people essential for the work, he still remained under NKGB surveillance. Vannikov said that Stalin was a very dangerous man. Stalin did allow people to defend their own point of view in discussions, one could argue with him. But once he had taken a decision, it was inadmissible not to obey it. The punishment meted out by Stalin for a failure to implement his decision was quick and cruel. It must be noted that the degree of severity of this punishment frequently was determined not by legal considerations but by Stalin's mood at that particular moment.

I would like to add another little story which illustrates Vannikov's personality. It is to do with the most eminent radio-physicist F.L. Mints, who was in charge of building the very first short-wave radio station in the USSR, named after the Comintern (on the Shukhov tower) and many others. He did a lot of work on the atom project. He had been duly elected as Corresponding Member of the Academy, but in spite of his having solved a great many problems, his election to full membership just was not forthcoming. He once had to travel to one of the PGU sites. Vannikov was bound for the same destination and he invited Mints to travel with him in his special coach. Over dinner Vannikov said to Mints:

"Well, another election has been held at the Academy of Sciences. And yet again they have not elected you?"

"Indeed they haven't."

"And you are upset about it?"

"Indeed I am."

"And those people who were voting — do you respect them?"

"No, of course I don't."

"So why are you upset, then?"

3.11. A.S. Yelian

Amo Sergeevich Yelian (born in 1903) was one of only three people who survived when on a night in 1915 Azerbaijanis attacked the town of Shusha in Armenia and murdered all its inhabitants. Shusha was for a time the second largest town in Armenia, a great many merchants lived there, Shusha was reputed to be a rich town. On that fateful date, Yelian was visiting a friend in a nearby village: this saved his life. After this massacre, no one lived in Shusha for 60 years: the Armenians were afraid and the Azerbaijanis felt that "there was a curse on that place". Yelian graduated from the Baku Polytechnical Institute. He spent almost a full year in the USA during 1936–37, on an official assignment to study American technology. In 1940 he was appointed Director of the artillery factory No. 92 in Gorky, which employed 45,000 people. The factory also had a large and most influential design bureau. At that time it was headed by V.G. Grabin, and later by Afrikantov.

During the Great Patriotic War this factory was almost the only one to work at full capacity: it produced 500 heavy guns every month. Stalin set a new target which demanded: a sharp rise in number of guns produced. Yelian knew that the factory's current equipment would allow it to raise its output by 5–10%, no more. As a result,

A.S. Yelian

its director — Yelian himself — would be shot. He analysed the situation and realised: each gun is planned to produce a certain number of shots: each shot produces micro-cracks in its barrel and eventually, when there are many cracks, the barrel explodes. Yelian decided to send engineers to the front to work out the life duration of a gun in battle conditions. They established that at the front a gun would fire 10 to 100 times fewer shots than assumed at the design stage. Yelian decided to simplify the production technology radically, reducing the strength of the barrel. But to achieve this, the factory had to be stopped completely for 5 or 6 days, producing no guns at all. This threatened the most serious consequences. But Yelian decided to take the risk — either of the choices open to him could end with him facing the firing squad, but one of them did give him a chance. Military representatives did not say anything for the first two days, but then they sent a telegram to the Kremlin: the factory had stopped the production of guns — sabotage. Yelian received an ominous telegram from Stalin, ordering him immediately to the Kremlin. But by that time the factory had managed to produce a gun using the new technology. Yelian had this gun loaded on a railway flat car and he himself traveled on that train. On arriving in Moscow, he managed to have the gun set up on Ivankovski square, right under the windows of Stalin's office. Once the gun was in place, Yelian went to Stalin and explained to him that henceforth, having introduced the new technology, the factory would be able to produce ten times more guns and therefore the loss of production that it had incurred would be compensated in less than one day. For this heroic action — and heroic it was — Yelian on 3rd June 1942 was awarded the title of "Hero of Socialist Labour" (he was one of the first six people to receive this award). Altogether, the factory produced over 100,000 heavy guns of various classes over the course of the war.

When the atom project started, factory 92 changed course: it started to produce the various equipment needed for this project: casings for nuclear reactors (both working on graphite and on heavy water), heat-producing elements, etc. Such a radical change-over was made possible thanks to the highly qualified and talented staff of the design bureau.

In 1955 Yelian was demoted from the post of Director of the factory: he was accused of wrong-doing in connection with the fact that his wife and Beria's wife were sisters. A.I. Alikhanov hired him to work at ITEP as a leading engineer.

Yelian and I used to spend our summers outside Moscow, our dachas were near neighbouring stations of the Kazan railway (Kratovo and 43rd Kilometer), so we often traveled to work on the same suburban train. That's when I heard the above story from him.

The People and their Attitude towards the Atom Project

———————————— • ————————————

4.1. Heisenberg's visit to Bohr in 1941

Heisenberg's visit to Bohr in Copenhagen at the end of September 1941 has been a topic of great interest for historians of physics. It also attracted the interest of the public at large after 1998, due to the production in London and later in New York of Michael Frayn's play "Copenhagen" — a psychological drama on the wartime meeting of two most eminent physicists who had been the closest of friends in earlier times: one of them, Niels Bohr, was living in Denmark which was at the time under German occupation and the other, Heisenberg, had come to visit him with the permission (or possibly even at the request) of the occupying power. In 2002 this play was also produced at the Moscow Arts Theatre. One cannot avoid asking some questions: what was the purpose of Heisenberg's visit? Was there an additional aim, apart from the obvious one of wishing to find out about the life on occupied territory of his old friend and teacher and that of his family, maybe to find out whether he could be given some assistance? And another question: what did they talk about? What did Heisenberg say (obviously, the initiative was his) and what did Bohr answer? All this is still being discussed even now, after the publication of Bohr's letters to Heisenberg about their meeting in 1941. These were written between 1957 and 1962 but they were never sent. The opinions expressed differ widely (see for instance [10, 37–41]).

I too would like to express my opinion on this historic and human tragedy.

First of all — the circumstances of the meeting. Nazi Germany was at the height of its success. The whole of Europe was occupied. On the Soviet-German front, the Soviet army had lost millions of combatants who were now held captive, and also a colossal quantity of war equipment: in particular, a great part of its tanks and aircraft had been destroyed. Kiev had been taken; Leningrad was under siege; the Germans were meeting only weak resistance on their progression towards Moscow. In Africa, Rommel's army was approaching the Suez canal; German protégés were trying to grab power in some countries of the Middle East (the Rashid Ali Gailani coup in Irak). Although the attempt to defeat Britain by air attacks had not succeeded, nonetheless the attacks by German submarines on transport from America were increasing. (I am presenting the situation in the way Germans in general and Heisenberg in particular saw it. If one looked from the other side, one might — if one tried — see some glimpses of hope for the anti-Hitler coalition.) It is possible that Heisenberg, who was well in with fairly high levels of Hitler's Reich, knew (or guessed) that Japan would soon join the war.

Now about the life philosophy of the two partners at this meeting. Heisenberg was a conservative, in fact one could say that he approved "strong hand politics" he was also a German nationalist — not a Nazi, not an anti-semite, but a nationalist who believed in the supremacy of German science and German culture and who was proud of it. The first of these opinions is based on the statement which he made to Dutch physicists during his visit to Holland in 1943 (I quote from [10]): "Democracy is incapable of ruling Europe. Therefore only one alternative choice remains: either Germany or Russia. In this light, the rule by Germany is the lesser evil".

During the Sudeten crisis in 1938 Heisenberg had been drafted to the army. When his unit was getting ready for the invasion of Czechoslovakia on the pretext of "freeing" the Sudeten area, Heisenberg did not experience any doubts in this respect.

And finally — presumably due to his pride in German physics — Heisenberg had come to the firm conclusion that since German physicists had not yet succeeded in making an atomic bomb, it meant that others were even less likely to succeed.[26]

Bohr was a man of entirely different views — he was a liberal of the Western type and a humanist. This latter is demonstrated by his behaviour after the war, when he wanted to stop the nascent nuclear arms race: he met Churchill and Roosevelt and endeavoured to persuade them to give the USSR information about the atomic bomb before it was used, in the hope that either Stalin would not create his own bomb, or that international control would be imposed straight away. This was naïve, of course, but this is precisely the sort of initiative a liberal would undertake.

These ideological contradictions between the two great physicists were not noticeable in prewar times, which were relatively tranquil: both were entirely devoted to the science that bonded and united them. They became manifest when the war unveiled all contradictions.

Heisenberg saw perfectly well that the Nazi leadership had no use for science as such, that it was being replaced by obscurantism and pseudo-science. He also thought that he — and only he — was able to save German science and to preserve it until the coming of future better times.

And lastly, a very important consideration: Heisenberg loved Bohr and had great respect for him. He knew that Bohr stood under serious threat. I think that by September 1941 Heisenberg already knew through his contacts in the highest spheres, or at least had guessed, the plans that were made for the "final solution of the Jewish question".[27]

---

[26]It is true that before the war German physics was indeed the most advanced. The theory of relativity and quantum mechanics were basically created in Germany, the principal physical reviews were published in the German language. But when the Nazis came to power in Germany, most of the prominent physicists left the country.

[27]The existence of Heisenberg's high-level contacts in Nazi Germany is demonstrated by several facts: (1) In 1937, an SS publication, the newspaper "Das Schwarze Korps" accused Heisenberg of teaching "Jewish physics" and called him "a white Jew". The Gestapo investigated Heisenberg's case for nearly a year. He was eventually declared innocent and loyal to the regime by Himmler's personal decision [10]. This shows that Heisenberg had seriously influential patrons protecting him; (2) In 1943 Heisenberg travelled to Krakow as a personal guest of the town's Governor-General Frank. This visit took place only a few months after the crushing of the Warsaw ghetto uprising [10];

(Göring's directive to Heidrich instructing him to prepare "the final solution of the Jewish question" was sent on 31st July 1941. The mass massacre of Jews on Polish territories had begun even earlier and had spread to the occupied region of the Soviet Union.) Bohr was half Jewish. He counted as Jewish by Nazi laws, which means that all directives concerning Jews applied to him. (In September 1941 these were not as yet extended to Denmark, this happened later. Approximately 500 Jews were massacred in Denmark.) Heisenberg wanted to save Bohr and Bohr's family.

Now for the state of the German atom project at the time of Heisenberg's visit to Bohr. The atom project was run by the Uranium Association (Uranverein), which was under the Research Branch of the Department of Armaments. Its head was Colonel E. Schumann. Several laboratories were included in the Uranium Association. In 1940, on Schumann's command, the group at the Kaiser Wilhelm Physics Institute of which Heisenberg was head became one of the Association's centres.

By September 1941 it was clear to German physicists that it was possible to establish a chain reaction based on slow neutrons in a uranium reactor with a heavy-water moderator. The reactor had to be heterogeneous — the uranium in the moderator must be placed in the form of plates or cylindrical slabs (Harteck 1939). The theory of the critical size of the reactor had been worked out by Heisenberg, and Flügge took account of the resonance absorption of neutrons by uranium-238. It was planned that such a reactor could be used as a source of energy. Later on in 1942, together with specialists of the Navy, the possibility of using reactors was considered for ship propulsion.

An attempt was made to separate uranium-235 (suitable to create an atomic bomb) from natural uranium, but it failed. In the summer of 1940, Weizsäcker (in Heisenberg's group) and independently

(3) Heisenberg counted among his friends: General Beck (executed in 1944 for taking part in a conspiracy against Hitler), also Ambassadors von Hassel and Count Schulenberg (see [10]). I know from experience in the Soviet Union that in the highest spheres secrets are not kept (apart from purely military ones), but rather they circulate in the guise of rumours — but they never get to spread downwards. I think that this was also the case in Germany.

Houtermans (in M. von Ardenne's group) showed theoretically on the basis of the Bohr-Wheeler liquid drop model that in a uranium reactor a new isotope of charge 94 and mass number 239 would be formed as a result of neutron absorption by uranium-238. Later this was called plutonium. Plutonium is fissile by thermal neutrons. The extraction of this isotope does not require isotope separation and could be carried out by much simpler chemical methods. This opened in principle the possibility of producing an atomic bomb. However, Heisenberg and other German physicists considered that in view of the huge work involved and the limited resources available, the atomic bomb could not be produced during the war and one had to concentrate on building a nuclear reactor [6].

In view of all that was said above it seems to me that one can imagine and understand the conversation Heisenberg had with Bohr in September 1941. According to the letters that Bohr wrote but never sent to Heisenberg, the latter had begun by saying that German victory was certain. Bohr, who lived under German occupation, could see what it had done to Denmark, and was hoping for a different outcome of the war, and could not accept such a statement. In view of the disparity between their situations, he could presumably not protest, but he could not agree either. Heisenberg said [6] that if the war were to drag on, atomic weapons would play a decisive part. He added that because he had been concentrating exclusively on this problem over the last two years, he knew that it was possible to create an atomic weapon: all the problems of principle had already been solved (see [6,10]). To support his statement, Heisenberg showed to Bohr a sketch of the nuclear reactor with uranium rods and control rods [10], the purpose of which, however, Bohr does not seem to have understood. Looking at the further development of the conversation, we may assume that Heisenberg had been overplaying his hand here: he said in so many words [6] that no work aimed at creating an atomic bomb was being carried out in Germany. Later, Heisenberg moved to the final purpose of the conversation, which is in my opinion the main one. In order to ensure the safety and survival of science and scientists in Germany, Denmark and other European countries, it was essential to make the Nazis realise that science was indispensable, and this aim can be attained if scientists helped Germany to win the war. The same thought was expressed in conversations with the staff of Bohr's institute by Weizsäcker who had accompanied Heisenberg on his visit. Heisenberg and Weizsäcker hoped that if the contribution of scientists to the victory proved to be substantial, this would ensure them a much stronger influence in postwar Germany, and this would bring about a softening of the regime. It was extremely important to Heisenberg for Bohr and his team to take part in the German atom project — this would have saved their lives. On the other hand, this would have also considerably strengthened the project. But for Bohr this line of thought and these propositions were entirely unacceptable, and he rejected them — presumably in fairly sharp words. Heisenberg lost heart and left.

Looking at all this now from a vast distance, one can understand why many years later Bohr, in his conversation with Feinberg in 1961, said that "Heisenberg had been a very honest man". This surprised Feinberg and he began to look for explanations to this apparent contradiction [10]. It is true that if one takes as starting point Heisenberg's belief in an inevitable victory of Germany, Heisenberg is shown to be truly honest. He saw no other means for saving German science or of saving Bohr's life. Moreover, Heisenberg's life philosophy did not lead him to see Nazism as the devil's evil — as we do — he rather saw it as some deviation in the history of Germany, which would eventually fade out so that life would return to normal again. (Heisenberg said [8]: "The Nazis ought to be allowed to stay in power for another 50 years or so, so that they would become completely decent human beings" — I quote [10,40].) It is quite possible (it is what Bethe thinks [41]) that Heisenberg did not want to create an atomic bomb. He postponed this problem, pushed it into the future and he possibly hoped that in the postwar world the scientists of Anglo-Saxon countries and Germany would voluntarily decide not to build it (Russia was not taken into account). The problem of creating the bomb did not inspire him as a physicist: in 1942–43 he did some excellent work on the S matrix, which had nothing whatever to do with the atom project.

Let us sum up. Heisenberg pursued several aims in his visit to

Denmark. From his point of view, these were all honest and generous. But from the point of view of Bohr (and ours), they were based on false premises, they were entirely unacceptable and even amoral.

Now about various hypotheses concerning Heisenberg's visit to Bohr. The first hypothesis is that Heisenberg visited Bohr on a spying mission: to find out what Britain and the US were doing on the atom project — thinking that Bohr might know something about it. (Or that this was one of the aims of this visit.) I consider this hypothesis to be entirely improbable. Even if Bohr had had some contacts with Western physicists (which in itself was hardly likely under the conditions imposed by the German occupation), they would hardly send him any secret information, for fear that it would get into German hands. Also, the very fact that this work was being carried out in Britain and the USA could be easily deduced from the fact that all mention of these themes had completely disappeared from scientific journals: in the USSR such a conclusion had been reached by Flerov.

Another hypothesis. Heisenberg wanted to use Bohr and his possible contacts in Britain and the US to reach an agreement with Western physicists for all sides to give up all work on creating an atomic bomb. (This thought is stated in Jungk's book [39].) This hypothesis considers Heisenberg to be much more naïve than he was. Let us assume that such an agreement could indeed have been reached. But what were the chances for its being adhered to, particularly under conditions of a world war? It is only now that a comparatively secure control has been established, because by now we have satellites and also a system of seismic stations monitoring underground explosions. But what did we have at that time? On the other hand, the possession of an atomic weapon would give any warring side a decisive advantage. And there would definitely be some scientists trying to create such a weapon. I do not think that Heisenberg was too stupid to understand this. In his letters Bohr unequivocally refutes suggestions that Heisenberg had proposed any of this to him or even hinted at it. I do not share Bethe's opinion [41] that Heisenberg showed Bohr the drawing of that reactor in order to persuade him that they (in Germany) were making reactors, not bombs. If Weizsäcker and

Houtermans, working on the basis of the Bohr-Wheeler theory, could come to the conclusion that plutonium is an explosive for the bomb and that it could be produced by using a reactor, then why could Bohr himself not arrive at this same conclusion?

4.2. Stalin and the hydrogen bomb

In our country, at least after the Revolution, science has always been strongly linked to politics. This connection became particularly strong in post-war times and it was strongest in physics, because physics was endeavouring to solve the fundamental problem which the State had set itself at the time: to create the atomic (and thermonuclear) bombs. This is not an exaggeration. The main purpose of the State (by "State" I obviously mean the ruling group at the very top) at the end of the forties and the beginning of the fifties was not so much to re-build the country's industry and agriculture which had suffered such terrible losses during the war, it was not even to reinforce the country's traditional military strength (this was in any case strong enough), but rather it was the creation of atomic weapons.

I am certain that Stalin aimed first and foremost to establish his world-wide power, or at the very least to take the first steps towards it — to annex Europe and a series of Asian territories (Turkey, Korea, access to southern seas — do remember the communist armies and the regions occupied by them in Greece, Indochina, Malaya, the Philippines, the blockade of West Berlin and much else). The aggression against South Korea was the first serious trial of strength. I had understood at the very start of military action there that this was an aggression carried out by North Korea and organised by Stalin, and that all the claims publicised by Soviet propaganda that it was South Korea who started or provoked the war were pure lies.[28]   I

---

[28] In connection with the Korean war, there was the following interesting episode. Academician Lev Artsimovich, the Vice-Director of the Institute of Atomic Energy, had a hobby — he liked to analyze war strategy, and he developed his own war strategy plans. At the time, when the South Korean and US armies were squeezed to the southern beach of the Korean peninsula, Artsimovich came to the conclusion that US forces would be set ashore from the Yellow Sea in the middle of Korea. He talked about this idea to some of his friends. Before long he was summoned to Beria who told him: "You know who is planning the operation?! Be silent, otherwise it will be bad for you". A few days later

also understood that this was Stalin's reconnaissance in force: if the West (and most importantly the US) had not rebuffed this attack, such actions would be repeated in other countries. I am convinced that in the early fifties Stalin intended to start and to win the third world war. Stalin had little time left — he had turned 70 in 1949 — and he had to act quickly.

Some important facts confirming this point of view have been made public recently. Lieutenant-General N.N. Ostroumov [42] who had been at the time Assistant Commander of Operational Directorate at the main headquarters of the Soviet Air Force, said in an article that in spring 1952 Stalin had ordered the creation of 100 divisions of new tactical bombers. In Ostroumov's opinion, this was a preparation for a new war. In Czechoslovakia the memoirs of General Chepichka appeared. Chepichka was Czechoslovakia's Minister of Defence in the Communist Government of Gottwald at the end of the forties–early fifties. In his book he relates among other things that in 1952 Stalin held a meeting of the Ministers of Defence of Socialist countries of Eastern Europe. At this meeting, Stalin declared that a world war was expected to take place within a year or two and he demanded of the assembled ministers that they prepare for it.

Two exceptionally difficult problems needed solutions if these aims were to be reached: a military one — the creation of atomic weapons — and a political one — to rouse the people to engage in war. The solution to the second problem was particularly difficult, and Stalin understood this perfectly well: to rouse the people to a new war only eight to ten years after the end of the harshest and most bloody war in Russian history, and moreover a war against a recent ally — America — would not be possible by using routine propaganda methods, even a campaign of terror was unlikely to succeed. He needed to rouse the fury by the people. But not an abstract fury against somebody far away behind the ocean, somebody who is known to the common man only through radio broadcasts. It was essential to make every man see right next to him the object of his

---

the US forces landed at Inchone, the North Korean army was defeated, and was restored only after the invasion of the Chinese People's Volunteer Army under the command of Marshal Peng Dehuai.

hatred, to understand that this enemy was threatening him and his entire family, and to know that these enemies were managed and instructed from a place beyond the ocean. It was not difficult to find a suitable object to be hated by the population — the Jews. Jews were ideally suited for this purpose: everybody had seen a Jew, everybody could have the object of his hatred right next to himself, and actually old Russian traditions of antisemitism had not as yet been forgotten. Starting in the second half of the forties, Stalin and his obedient Party apparatus were purposefully fanning the flames of antisemitism.

A campaign was started, aimed at fighting cosmopolitanism and excessive veneration of anything foreign. (Scientists who liked to joke defined the tangent of the angle of excessive veneration as the ratio of the number of quotations of foreign articles to that of quotations of Soviet ones.) In arts, people were losing their jobs and their positions, written material did not get published (this applied to both writers and journalists), Jews experienced difficulties in their attempts to gain admission to universities, and it became practically impossible for a Jew to get to study at the Mathematics Faculty of Moscow University. The campaign was particularly fierce against professional critics in literature and music. This was conducted personally by A.A. Zhdanov, Secretary to the CPSU Central Committee. The antisemitic campaign rose to another level with the murder in 1948 of S.M. Mikhoels, a remarkable actor of worldwide fame, the creator of the Jewish Theatre in Moscow. During the war, Mikhoels had traveled to the US and had collected a great many donations for the Red Army there. His murder was organised by agents of NKVD: first, Mikhoels was killed, and then his body was run over by a lorry. A medical commission led by Academician Burdenko concluded that the death of Mikhoels was the result of a traffic accident. (The same man — N.N. Burdenko — headed the commission on the death of 20,000 Polish officers, prisoners of the Red Army. Burdenko's commission concluded that they had been killed by Hitler's forces. In actual fact — as is known and generally accepted now — they had been killed by the NKVD.)

The antisemitic campaign which had continued to grow right

up to the moment of Stalin's death was not just yet another episode in Stalin's policy of repression against whole nationalities that
displeased him — it was a means to a far-reaching end. A new and
very important stage on the road towards this end was the "Affair of
the Doctors". At the end of 1952, a group of doctors was arrested,
all of them professors — the best specialists in their own fields. All
of them (with the exception of one or two) were Jewish. They were
accused of acting on the instructions of the American Jewish spying
organisation "Joint", attempting to kill leaders of the party and of
the State while pretending to give them medical treatment. From
the very first announcement on the "Affair of the Doctors" it was
clear to me that this was a fabrication on Stalin's instructions and
that this was the start of a new campaign. Unfortunately, the fact
that the "Affair of the Doctors" was a total fabrication was not clear
to quite a few people, even among the best educated: very many
were taken in. The "Affair of the Doctors" had been devised to produce a long-range effect: it was meant to prove that among the Jews
even people of the noblest profession — medical doctors — were in
fact murderers. This was not restricted to a couple of dozen eminent
doctors who had been arrested and imprisoned: rumours were spread
all over the country that all Jewish doctors were enemies of the people and criminals. I personally have heard repeatedly in the street,
in shops and at other similar circumstances statements of the sort:
"The doctor in our local health centre is a Jew. I won't go to him,
he will poison me" or "So-and-so has died in hospital, he was killed
by the doctor — a Jew". This hatred spread further and further and
eventually it was also directed at others, not only at doctors.

The scenario was to develop thus: The people arrested in the
"Affair of the Doctors" were to be executed in public. At the same
time it was planned to have "spontaneous" outbursts of the population against Jews. And then a group of prominent representatives
of the Jewish people was to send a letter to Stalin and the Soviet
government in which the authors would recognise the collective responsibility of the Jews as a nation for the fact that such monsters
had been allowed to grow in their midst, and the just anger of the

people would be justified.[29] The authors also would have asked to
protect the Jews from the people's wrath — to ensure their safety
they would be moved to designated regions of the Soviet Far East.
The camps destined to house them were already being prepared,
some were being built. According to this plan, spontaneous mass
demonstrations would take place along the route of the transport
taking the Jews to this destination. It is easy to foretell a strong
reaction from America: America would of course rise in defence of
Jews. Western Europe would support America. And then — according to Stalin's plan — it would be possible to switch the people's
anger from the enemy within to the enemy abroad.

A second problem had to be solved as well, the military one. At
the end of the 1940s, the Soviet Union had an undisputed supremacy
in Europe as far as land armies were concerned. But this was not
enough: even if parity with the US in atomic weapons could not be
achieved, it was essential to produce at least so many of these and of
such quality that the Americans would have cause to fear an atomic
strike against the US, and to give serious thought before using atomic
bombs, should a new war erupt in Europe.

As from 1949, the USSR already had atomic weapons. But only
in small quantity, and in this respect we were lagging far behind
America. In 1945 it became known that there was important work
being done in the US, aimed at creating a much more powerful type
of weapon: a hydrogen bomb. This work was at the time still far from
completion. In the same year the idea of creating a hydrogen bomb in
the USSR was put forward by some physicists: I.I. Gurevich, Ya.B.
Zel'dovich, I.Ya. Pomeranchuk and Yu.B. Khariton. At that time
this idea was not developed. In 1949 it was decided to intensify the
effort on the creation of the hydrogen bomb with a genuine chance
of catching up with America. Some of the groups co-opted for this

---

[29]According to information in my possession, such a letter already existed. It had been
written by Academician Mints, (a specialist in the history of the CPSU), moreover the
letter had already been signed by some personalities. I know the names of at least two
people who signed it — of course, they had been under extreme duress. These people
are long since dead and I do not wish to disturb their memory, so I shall not name
them. I shall however name one courageous man who refused to sign: I.G. Ehrenburg.
(According to other sources, the singer and People's Artist of the USSR, M. Reizen, and
General Ya. Kreizer also refused to sign this letter.)

had never done any work whatsoever on the bomb, others had only dealt with separate problems connected to this project. (Among the groups co-opted were those of I.E. Tamm, which included A.D. Sakharov, the groups of N.N. Bogoliubov, of I.Ya. Pomeranchuk and others.)

I should like to stress that in my opinion the aim was not to win a nuclear war against America after having overtaken the US in the race to create a hydrogen bomb. I think that Stalin did understand that this was impossible. The aim was different: to create a hydrogen bomb at roughly the same time as the American one, to test it and to demonstrate that we too were in possession of nuclear arms. The Americans would not know how many nuclear bombs we would have — two, three, or five. And if there was to be a war in Europe in which only conventional weapons were used (of course this would indeed be a blitzkrieg in view of the marked superiority of the USSR's terrestrial forces), it was very probable that the US would not use nuclear arms, for fear of a nuclear strike onto their own territory. A Soviet hydrogen bomb would thereby be a means of nuclear blackmail if such a war started in Europe.

The way in which events developed fully confirms this scenario. At the end of 1952 it became clear that the hydrogen bomb would be created very soon (within six months to one year): all problems of principle had been solved, basically only technical solutions still needed to be found. The middle of 1950 saw the start of designing and then constructing reactors for the production of tritium, the main essential component of the hydrogen bomb. Political preparations were developing at the same time: in December 1952 — the "Affair of the Doctors" was launched, its conclusion could be expected in the spring or summer of 1953. The testing of the hydrogen bomb in the USSR was carried out in August 1953; it may very possibly have been delayed by Stalin's death and the perturbations that followed it (the execution of Beria, the changes of leadership in the atomic industry, etc). I am therefore deeply certain that if Fate had not intervened — Stalin's death in March 1953 — the third world war would have broken out in 1953 or 1954, and the world would have found itself on the verge of catastrophe (or even beyond). Therefore from my point of view the creation of a hydrogen bomb in the USSR in the early 1950s would have been the most terrible danger for mankind.

4.3. The attitude of some physicists towards the atom project

As far as British and American physicists were concerned, their situation (and duty) was crystal clear, once it had become obvious that more than two neutrons were born in the process of division after the capture of uranium by the nucleus, in other words — once it had become clear that a chain reaction was possible. Germany was ruled by Hitler; people who fought Fascism and people who were Jewish were being sent to concentration camps, therefore it was utterly unacceptable to allow the Nazis to become the first to own nuclear weapons. Therefore practically all physicists who had some connection with the atom project — be they British or American, not to mention those who had emigrated from Germany — devoted themselves unreservedly to work on the atom project. Only a few major physicists remained in Germany: Lenard and Stark who were active Nazis, Laue — who did not leave Germany but was nonetheless not a Nazi, and Heisenberg, who was the head of the German atom project, but hoped that "in fifty years the Nazis will become decent people" (these are Heisenberg's own words[30]).

The situation changed once Germany had been defeated. Many physicists felt that the danger was past and that they could return to their interrupted research and teaching. Oppenheimer left his post of Director at Los Alamos, Bethe returned to Cornell University. However, when the Soviet Union exploded its first atomic bomb, a change arose in the attitudes of many physicists and more importantly those of politicians. The Cold War was raging, the Blockade of Berlin started in June 1948, the People's Republic of China was formed in 1949, Communist forces advanced in Greece and Vietnam, the *Pravda* newspaper published an article detailing claims to the Turkish provinces of Kars and Ardagan. In these circumstances Truman took the decision to continue the research aimed at creating

---

[30]I am quoting from [10, 39].

the hydrogen bomb. Many Western physicists — including Oppenheimer, Compton and several others — however did not support this decision. Bohr, doing his own thing, attempted to persuade Western leaders to conclude a treaty with the USSR which would prohibit nuclear arms, but he did not succeed either in the West or in the East.

Here I come to the delicate question of the part Soviet physicists played in the creation of the hydrogen bomb. It is not pleasant to say this, but I must: the overwhelming majority of the eminent physicists of my acquaintance who were in some way involved with this problem (but not all!) were not aware of this menacing danger — on the contrary, they were certain that the creation of atomic and thermonuclear bombs in the USSR would help to prevent a new war and that it would be a defence against a possible American aggression. Therefore they worked as well as they could, they showed initiative, they did not spare their energies or their time.

The atomic bomb in the USSR was created in 1949. It is however openly admitted now (including by Khariton who was at the head of this programme) that at the start we followed the path laid by the Americans, since we had at our disposal data about the design of their atom bomb. A totally different situation arose concerning the hydrogen bomb. The Soviet hydrogen bomb was entirely original, and the credit for this is due to A.D. Sakharov. As is well known, in the hydrogen bomb the reaction of the fusion of tritium T and deuterium D+D and D+T takes place. In the late forties and early fifties, when the question of the creation of the hydrogen bomb arose, the USSR had practically no tritium. (Tritium is unstable, its half-life is 12 years, and in nature it exists in negligible quantities.) Tritium can be produced in nuclear reactors working on enriched uranium. In the early 1950s, there were no such reactors in the USSR and the decision to create them had only just been taken. It became obvious that it would not be possible to produce a significant amount of tritium over a short time (2–3 years). But Stalin was urging the work on. (I could of course not know this directly, but I could deduce all this from studying the development of work on reactors for the production of tritium — I was involved in this work.) Therefore it was extremely important to develop a type of hydrogen bomb which would necessitate only a minimal amount of tritium. This is the problem which Sakharov solved. He thought up — yes, he literally thought this up, it was his own idea entirely — a way to construct a hydrogen bomb with a minimal quantity of tritium. Here I can quote the words of Pomeranchuk who told me one day: "Sakharov is not so much a theoretical physicist — he is an inventor of genius!" At the time I did not know the nature of Sakharov's idea (in his "Memoirs" it is called idea Number One). When Pomeranchuk talked to me about it, he only said the word "sloyka", leaving me to guess everything by myself.

The nature of this idea is now known. This is the one that allowed to explode the first hydrogen bomb in the USSR almost at the same time as the American one. (The first testing of the American hydrogen bomb was carried out on 1st November 1952. This bomb was different from the Soviet one in that it was too heavy to be transported — therefore it could not be used as a weapon. The first American hydrogen bomb capable of being transported was tested half a year after the Soviet one.)

Sakharov's idea "Number One" consisted in creating alternating layers of deuterium and tritium. When the fast neutrons ejected by the tritium split the deuterium, they produce additional neutrons. Soon after, Ginzburg [11] supplemented this idea by proposing to replace the deuterium by LiD, which greatly increased the neutron flux. (Sakharov's Idea Number Two [5].) The third idea, according to Sakharov, was to surround the explosive mixture with a heavy casing which largely reflects the γ quanta and neutrons, thus creating pressure in the explosive mixture, and as a result the temperature increases, and at such a temperature, if uranium 238 is put in the explosive zone, then the uranium-238 is fissioned by the fast neutrons, i.e. the energy of the explosion is greatly increased. In this way an unheard-of energy of 100 MT is achieved.

Towards the end of 1952, Stalin already knew that the work on creating a hydrogen bomb for us was heading for success and I believe that this fully correlated with his political plans. As I understand it now, the actions of scientists who were giving their all to create the hydrogen bomb also had a negative aspect to them. Here I would like to comment on what I was saying earlier: among the scientists who were working on the atom project, not everybody behaved thus, not everybody was so blind. The exception was L.D. Landau. This is shown in a brief remark of his mentioned in Sakharov's "Memoirs". I shall quote it word for word, because it is very important.

"Once in the mid-1950s I (Sakharov) came on some errand to the Institute of Physical Problems where Landau headed both the Theory Department and a separate group that was carrying out research and calculations for "the problem". Having completed our business conversation, Landau and I went out into the Institute's garden. This was the only time that we spoke without witnesses, quite confidentially. Landau said:
– I really do not like all this. (From the context I knew that he meant nuclear arms in general and his participation in this work in particular);
– Why? — I asked naïvely;
– Too much noise.

As a rule, Landau smiled often and readily. But on that occasion he was sad, even sorrowful.

This brief conversation shows the whole of Landau as a person and of his attitude to "the problem". His last reply is particularly characteristic. Landau stuck to the following rule: if a man does not immediately understand something which is self-evident in Landau's own opinion, there is no point in giving explanations — the conversation must be cut off, and Landau would do this by uttering some sentence of little importance.

Landau did work on "the problem" and he did work on it conscientiously, and this to his own scale of conscientiousness. He carried out all the tasks that were set for him to the highest possible degree so that no one could quarrel with his work. But he did not show any initiative and he tried to step aside whenever possible. The greatest caution had to be exercised here, one could easily pay with one's life.

When the news reached Landau that Stalin had died, he rushed out of his room, dancing on the spot and exclaiming "He's croaked, croaked, croaked! Now I am no longer a slave, I won't work on the atomic bomb any more!" V.L. Ginzburg relates that he once asked Landau: "What would you have done if the idea of the hydrogen bomb came into your head?" Landau replied: "I would not have been able to resist the temptation of doing all the calculations, I would do them and then I would tear my notes to bits and flush them down the toilet".

Zel'dovich's attitude also fluctuated. Zel'dovich came from an educated family and he was a first-class physicist. It seemed that he was able to calculate ahead of the event the consequences of the tests of the RDS-37 bomb held in 1955 (a two-year old girl and a soldier died) and to foresee the danger of the consequences spreading all over the Earth. Nonetheless, when the tests were successful, he ran up to Sakharov and hugged him with shouts of joy "It worked! It worked! Everything worked!" As to Sakharov himself, in 1955 he was experiencing "a range of contradictory feelings, and the main one was the fear that the energy thus liberated might get out of control and bring incalculable calamities".

Shortly before his death Zel'dovich said something in an entirely different key. Talking about his three decorations of "Hero of Socialist Labour", he said: "For these three Stars I sold my immortal soul to the devil".

Sakharov writes in his "Memoirs": "What we were doing was in fact a great tragedy which reflected the whole tragic situation in the world where for keeping the peace it is essential to do such dreadful, appalling things. Today, thermonuclear weapons have never once been used against people in war. My most passionate hope (deeper than anything else) is that this may never happen, that thermonuclear weapons will contain war but will never be used".

In 1955, Kurchatov, Alikhanov, Kikoin and Vinogradov wrote an article in which they analysed the possible consequences of an atomic war and stated in conclusion: "Humanity is now overshadowed by the immense threat: the end of all life on Earth" [7,14]. Up to that time, Soviet propaganda had claimed that a new world war would mean the end of the capitalist system. The article was also signed by the Minister for Medium Machine Building V.A.Malyshev. It was sent to Malenkov, Khrushchev and Molotov. Malenkov appears to have shared the opinion of the article's authors: he did say in one of his speeches that a new world war would bring the destruction of world civilisation. Khrushchev, however, rejected this opinion, calling it "theoretically incorrect, erroneous, and politically harmful". The Party returned to its old formula and the article was not published.

Kurchatov had a highly developed sense of responsibility. A.P. Aleksandrov testified that after the test of the hydrogen bomb in 1955 Kurchatov came back in a state of deep depression and told him: "...what a terrifying thing we did. There is only one path left for us to concentrate on — this is to ban this stuff and to exclude the possibility of a nuclear war".

I shall now write about the hydrogen bomb project in which I took part myself. Work on this started with a proposal made in 1945 by Gurevich, Zel'dovich, Pomeranchuk and Khariton — I have already written about it.[31]

Their idea was the following (this system was nicknamed "the tube"). A long cylinder would be filled with deuterium. A tritium fuse was placed at one end of the tube — this would be fired in one way or another and would create a very high temperature. Consequently a blast wave caused by the reaction D + D would travel along the tube. This system could be built to any length and it was cheap to make, because deuterium was inexpensive and one would need tritium only for the fuse. For this type of bomb, the strength of the explosion was restricted only by the resources needed for its transport. For instance, there was an idea under discussion to bring this bomb (suitably camouflaged) to the shores of America by ship and to explode it there, which would destroy the whole coast (in his "Memoirs" [5] Sakharov mentions his discussion of a similar idea with Rear-Admiral Fomin. Fomin's response is interesting — its gist is: "We seamen do not wage war on the civil population").

Until recently I thought that the proposal put forward by Gurevich *et al.* had been original. Gurevich himself was sure of it himself. It is now known, however, that a similar project was being developed in the US — there it was called "classical Super". Its concept

---

[31]Cf. Gurevich I.I., Zel'dovich Ya.B., Pomeranchuk I.Ya., Khariton Yu.D. Report of Lab. No 2. - IAE, 1946, published in UFN, 161, 171 (1997).

was formulated as far back as 1942 by E. Fermi in a conversation with E. Teller, the future "father of the American hydrogen bomb". Teller started to develop this idea and worked on it intensely over several years. In spring 1945 Soviet intelligence reported the first information about the American hydrogen bomb project, and more detailed data about it were reported in October 1945. Gurevich and the others made their presentation on 17 December 1945.[32] This is why I think that the idea of the Soviet "tube" was conceived from intelligence data. But the concrete and detailed working-out of the project is undoubtedly original. I am entirely sure of this because I knew two of the authors (Gurevich and Pomeranchuk) very well indeed: they could not possibly appropriate somebody else's idea, passing it as their own. Why then did Gurevich think that everything in their presentation, including the main idea, was original? Well, information received via intelligence was passed on only to a very restricted group of people: only to Kurchatov, Khariton and possibly Zel'dovich, and these people could not quote the source of these ideas when they presented them to others, they were obliged to pass American ideas as their own. This is why Gurevich (and apparently also Pomeranchuk) sincerely believed that all the presentations were the work of four authors from beginning to end.

As far as I know, there was no work done on this project until 1949 — presumably because the atomic bomb had not been created as yet and all efforts were directed at that target. Moreover, one needed a fuse to ignite the tritium. This fuse had to be an atomic bomb. If one did not have an atomic bomb, there was no point in starting detailed work on a hydrogen bomb. Detailed theoretical calculations for the "tube" started in 1949 or 1950, they were mostly done by Zel'dovich's group in Arzamas-16. Landau's group took part in this work too, but it was solving only individual problems, separated out from the overall problem. The main problem, whose solution would determine whether or not it was possible to create such a bomb, consisted in finding out what the energy balance would be. In order to produce a self-sustaining nuclear reaction (that is in

---

[32]The author is grateful to G.A. Goncharov for this information. See also Goncharov G.A., *Thermonuclear Milestones*, Physics Today **49**, November 1996, 44–61.

order to make the bomb explode) it is essential that this balance be positive, in other words it is essential that the energy produced by the nuclear reaction be greater than the energy which escapes from the system. The Zel'dovich group did the calculations for the "tube" and got the result that the energy balance was zero, in other words, the energy produced by the nuclear reactions is equal to the energy that escapes. However, the accuracy of the calculations was rather low, somewhere in the region of a factor of 1.5 to 2. If this factor worked towards a positive energy balance, the bomb could be made. If the factor was negative, the bomb would not explode — at the time people said "we would get a 'pshick'". The latter outcome was of course unwelcome to everybody. This style of calculations — with this degree of accuracy "up to two" — was actually typical for Zel'dovich. It was very good in some cases and did sometimes lead to striking results, but in this particular case it did not work.

Entirely different methods were needed to achieve greater accuracy, to bring it to 10–20% accuracy. Zel'dovich's group could not cope with this task on its own. This was in the mid-fifties. Pomeranchuk was then ordered by senior management to take a lengthy business trip to Arzamas-16. Pomeranchuk was very unhappy there and suggested to the management that he would undertake the task of solving this problem in collaboration with his group at the TTL, on condition that he would be recalled back from that base.

Pomeranchuk's offer was accepted, he returned to Moscow and sent up a list of names of his people to be included in the group. He had to do this because, although we all already had very high levels of clearance concerning secret information (since we were all working on reactors), this particular task required a very special clearance, at an exceptionally high level: all documents on this topic were classified "under 4 letters" ("s.s.o.p." — the Russian initials of the words "top secret, special file"). The main reports in this category were written by hand, they could not even be entrusted to a typist, even one with the highest possible clearance. (Pomeranchuk wrote the final report himself by hand, in three copies, without using carbon paper.) The physicists included were B.V. Berestetsky, A.D. Galanin, A.P. Rudik and myself. The mathematical part was headed by

A.S. Kronrod. The mathematical calculations in this problem were difficult and had great importance. Kronrod was happy to take on this sort of problem — it was a sort of challenge for him. And in fact he did invent an effective method for the numerical solution. In those days there were no computers, calculation aids consisted of manual electric calculators. M.V. Keldysh was the head of the commission in charge of providing mathematical servicing to the atom problem. He allocated to the problem a most powerful computing bureau, headed by L.V. Kantorovich, a future Nobel laureate. This bureau was situated in Leningrad, it had a staff of approximately forty calculation technicians. While working on this problem, Kronrod put up a first-class performance, much better than that of Kantorovich. I was frequently present at their discussions: the ideas were always put forward by Kronrod, and Kantorovich only implemented them. This may be due to the fact that Kantorovich was definitely not enthusiastic about the problem under consideration. (Although the data were presented to him in such a way that their physics was hidden (with a clearance merely "under two letters") but — I think — he did guess what he was doing.)

Among the physicists, Galanin did not take part in this problem at all, since he was entirely taken up by work on the reactor. Berestetsky was solving specific problems connected with it. Therefore only Rudik and I started on it. Initially we needed to check the report by Landau, Lifshits, Khalatnikov and Diakov which contained the calculations of the cross section of Compton scattering on electrons in plasma. As we were checking this, we discovered that the calculations were not correct. And something unexpected happened then. We had started to work without waiting for the official clearance for it — the work could not suffer any delay. Clearance was granted to everybody — apart from Rudik. Aleksei Petrovich Rudik, a descendant of Cossacks, who was at that time Secretary of the Komsomol organisation of TTL, did not get clearance while I, Ioffe Boris Lazarevich, who was not a member of the Party, who had never even been a member of Komsomol — did get clearance! There were indeed grounds to feel amazed. As a result, I was the only one from our group of physicists left to do the work. Pomeranchuk did take part in discussing results, particularly at the final stage, but he did not do any tangible work. The calculations were completed at the end of 1952. The result showed that the energy balance was negative, in other words, if we take as our unit the energy produced in nuclear reactions, then the energy escaping from the tube was 1.2. The system did not work, such a bomb could not be made as a matter of principle. This was very fortunate for mankind — or perhaps God took mercy on it.

Now I would like to discuss the attitude of different people to "the tube". First — Alikhanov. The work was done in TTL (Alikhanov was Director of TTL) and it could by no means by-pass the Director, since all significant projects (and this was a very significant one at that time) had to be known to the Director. Alikhanov, however, adopted a very clear position right from the beginning: "You want to do this work — you carry on, but I don't and won't have anything to do with it". He issued the order that all papers concerning this project would be signed by Pomeranchuk, by-passing him, Alikhanov, and he distanced himself from this activity right to the very end, when the final report containing a negative result had to be signed.

Landau did take part in the initial stage of the research on this problem, but then he distanced himself. At the end, when it became clear that the system did not function (the energy balance being negative), then — the energy balance was only weakly negative — the question arose whether one might find some physical effects, hitherto not taken into account, which might improve the balance or somehow change the system for the same purpose. These questions were discussed many times in 1952–1953. Their participants were people belonging to Pomeranchuk's and Zel'dovich's groups and also B.B. Kadomtsev and Yu.P. Raiser from Obninsk. They were studying a similar system — "the sphere". Although with that system it had been clear right from the start that too much tritium was needed and did not allow the effect as hoped to achieve with "the tube" (an unlimited energy of the explosion), the sphere proved to have much in common with the tube from the point of view of theoretical calculations. Landau was also invited to these discussions. When he was asked whether some effect or other might affect and change the situation, his answer was always the same: "I do not think that this effect could be of material significance". Once it became clear that "the tube" would not work, Pomeranchuk said that he had no ideas on how to improve the system and that he could no longer continue this work. He offered to take me on the study of some questions which remained somewhat unclear and added that he would organize my being appointed as head of the group which would carry out this research. But I refused, saying that I had no ideas either. As no one was found who wished to continue, this problem was closed.

Landau's position in this had been very important. When he said that he did not think that the effect mentioned could have any material significance, this caused even those people who initially had desired to carry on the relevant calculations to change their mind and lose this wish. A similar position was adopted also by E.M. Lifshits — he always endeavored to remain uninvolved, to refrain from showing any initiative of his own.

In the US, once the atomic bomb had been created and once the war was over, many physicists began to doubt that it was essential to continue working on the atom problem, particularly on the creation of a hydrogen bomb. Many scientists returned to their universities, to resume the science research and the teaching which had been interrupted by the war. Many were of the opinion that the US did not need to create the hydrogen bomb, that this would even be harmful for the country. The discussion on this matter between R. Oppenheimer and E. Teller as well as the ensuing "Oppenheimer affair" are well known.

Nothing of the sort happened in the USSR. One may well ask — why not? The obvious answer is that people were afraid, but this cannot fully satisfy us. The situation is not made any clearer by referring to the habit, deeply rooted in the Soviet man, to obey commands without any thought about it — as the song says: "But if something goes wrong, that's none of our business, we just did do what our Motherland told us to do". If the work carried out by scientists on the atom problem could be explained entirely by their having been forced to do it, the results could not have been achieved, particularly in such a brief time. This sort of work needs creativity, it needs initiative, which is impossible in forced labour conditions. And lastly, the explanation "this is superb physics" (to quote Fermi) is not satisfactory either, because it is valid for the physicists of the USA and USSR in equal measure. It seems to me that the explanation lies in the fact that the majority of those who created the hydrogen bomb were people who belonged to the 1930s generation, people who (at least to a degree, some more and some less) believed in Socialism and believed that it could be built in the USSR. It is only slowly, frequently at the cost of a most painful revaluation of their convictions, that they came to realize the truth — that the dreadful weapons which they were creating were to be handed over to out-and-out criminals. In this respect, the memoirs of Sakharov, written with the greatest sincerity, are very typical: they show that for him, this realization began to emerge only in the 1960s. (This did happen earlier to some other people, though.) Such a delay in changing opinions was not restricted to scientists only. It affected writers, poets, artists to an even greater extent. Let us remember Gorky's words: "If the enemy does not surrender, he must be exterminated!" or paraphrase Mayakovsky — "Machine-gun down those fleeing scoundrels!" These two writers are not alone, there are very many more people who are entirely decent (by our modern standards) members of the writing and thinking class who nonetheless are the authors of some statements which make one deeply wonder — how on earth can one say or write this? But very rare are those who succeeded in maintaining the clarity of their thoughts, the honesty of their actions and of their judgements.

A Look to the Future

5.1. Power engineering

Everybody — an economist or a geologist alike — says that the Earth's oil and gas reserves will run out in 50–100 years' time. Reserves of coal are greater, but it may be impossible to use them because the burning of coal causes unacceptable atmospheric pollution. Energy can be produced in nuclear fusion power stations, but their contribution will hardly be noticeable until 2050, moreover, since we have no reliable experience in this field, no one knows what unexpected problems may arise with their use. It is difficult to imagine that solar energy, wind energy and the energy obtained by processing biomass will be sufficient to provide for the work of power stations: these are estimated to need 5,000 GW p.a.

There still is nuclear energy production. But here we encounter two fundamental problems: the safe disposal of waste products and the non-proliferation of nuclear weapons. Let us assume that the first problem will somehow be solved. Eight countries already have nuclear arms. Let us assume that they will somehow come to an agreement and that these arms will indeed be under control. (There still remains the strong possibility that some group of terrorists will succeed in capturing some of these arms and use them, or else that some mad dictator with access to an atomic bomb will appear in some country.) But many more countries possess nuclear power stations equipped with reactors which produce large quantities of plutonium. It was previously assumed that the plutonium produced by a power station cannot be used to make an atomic bomb, since it contains a sizeable admixture of Pu-240, which has a high probability of spontaneous fission. (Two pieces of such plutonium will heat up before a nuclear chain reaction can start, so they will simply fly apart.) But it has nowadays become clear that a bomb can be made when PU-240 is present, whatever its quantity. A nuclear power station of 1 GW of electric power produces approximately 200 kg of plutonium per annum. Such quantities are difficult to keep under control, and it would be enough to steal just 2–3% of this to make a bomb.

Fiveson and Taylor [45] and Bethe [41] proposed an idea concerning nuclear power reactors which would ensure the non-proliferation of nuclear arms by the force of objective laws of physics. For this, reactors must be of two types. In reactors of the first type, the fuel rods contain ^{238}U, ^{232}Th and ^{235}U; during operation ^{235}U gets replaced by ^{233}U, which is produced from thorium. The moderator in such reactors is heavy water. One can expect a high (near 1) conversion coefficient, i.e. the ratio of the fission elements formed (primarily ^{233}U but also ^{239}Pu and ^{241}Pu) to the burnt ones. The ratio of the concentration of ^{238}U, ^{235}U, ^{232}Th is chosen so as to make sure that at every stage of the reactor's work (including at the unloading of the fuel), the nuclear fuel is unusable for making an atomic bomb. (At the initial stage, when the fuel does not contain ^{233}U, the content of ^{235}U is less than 20%, and ^{235}U$/\left(^{238}$U$+^{235}$U$\right)$ is also less than 20%; at full replacement of ^{235}U by ^{233}U — which stage is not reached in reality — the content of ^{233}U is less than 12% — as demanded by IAEA.) The design of his type of reactor can be based on the design of the Canadian CANDU reactors or of the Czechoslovak power station described in Chapter 2.13. The second type of reactor is of the breeder type working on ^{238}U and plutonium. The reflector of a breeder must contain some thorium in which ^{233}U is produced, thereby transforming the breeder back into a reactor of the first type. The number of breeder reactors needed will be much less than that of the first type reactors, since the latter are almost self-supporting. Breeders must be sited in countries of high stability and must be subject to international controls. Only reactors of the first type may be the object of international business. Of course, this pattern of nuclear power stations demands the establishment of international agreements, but this is the price to pay for international stability, i.e. the price we must pay for peace. Definite plans for such power stations have been worked out in calculations by Radkovski and Galperin [47], Ponomarev-Stepnoi [48] and by Ioffe and Kochurov [49]. As a result of the calculations in [49] it was established that in respect to the spent fuel, the ratio of fissile uranium isotopes $^{235}U + ^{233}U$ to the entire quantity of uranium is 11.2%, which is less than the IAEA norm. Also the annual production of plutonium is only 17 kg. Therefore, to produce a bomb, one would need to steal one third to one half of the yearly production of plutonium, which would be easily noticed.

5.2. Is the world cognizable?

Theoretical physicists have been longing for years for the creation of a unified theory of elementary particles, of a theory which would unite all interactions — strong, electromagnetic, weak and gravitational. 30 years ago, success in formulating the string theory of gauge fields gave reasons to hope that the path towards creating such a theory was open. But then it became obvious that the number of versions of string theories is unbelievably large (by some estimates, their number is 10^{500} and even possibly 10^{1000}) and there is no way of choosing the true one. Even worse, over these 30 years of string theory, no one could produce even a single physical prediction. These considerations lead me to believe that this particular path towards the creation of a unified theory is closed.

High energies up to those of the Large Hadron Collider (LHC) are now accessible to experiment, that is energies up to 14 TeV which correspond to small distances of 10^{-18} cm. The study of the region of small distances above 10^{-18} cm have confirmed the Standard Model, which includes weak and electromagnetic interactions (the electroweak theory) and strong interactions. The Higgs boson with mass 125 GeV, predicted by the Standard Model, has been discovered.

Thanks to the discovery of the Higgs boson, the electroweak theory is now complete, uncontroversial and consistent.

Gravitational interactions stand apart. Their characteristic distances, that is distances at which the gravitational interaction becomes strong, is of the order of 10^{-33} cm. Experiments will never be able to advance through the area from 10^{-18} to 10^{-33} cm. The simple fact is that the Earth's resources are not sufficient to build an accelerator for energies which correspond to the distance of 10^{-33} cm: this would need energies of $\sim 10^{16}$ TeV. (In fact, the energy consumed by LHC is of the order of 100 MW. To increase the energy of an accelerator up to 10^{16} TeV a power of $\sim 10^{17}$ MW $= 10^{11}$ TW would be needed, whereas the present overall output of all power stations in the world is of the order of 2–3 TW.) The region of distances from 10^{-18} to 10^{-33} cm is not empty: this is demonstrated by the fact that neutrinos do have mass and oscillate, and also by the existence of dark matter, neither of which facts fits into the Standard Model. There is no reason to believe that in the energy region from 10 TeV to 10^{16} TeV there truly is nothing at all apart from the interactions known to us. Since this area will never be reached experimentally (with the exception of its lowest edge), we come to the conclusion that the area of small distances is not cognizable to the physics of elementary particles.

Another example concerns the entropy problem. According to the second law of thermodynamics, in a closed system all processes take place in such a manner that with the course of time entropy either rises or remains constant. This fact determines the direction of the time arrow. The growth of entropy with time is not linked to the structure of the Lagrangian, since the Lagrangians of all interactions (except the weak interaction) are invariant under time reversal and the weak interaction is immaterial in macroscopic processes. Let us go back in time, approaching the moment of the Big Bang. Weinberg [50] has shown that on the basis of known laws (which are symmetrical relative to the exchange $t \to -t$) it is possible to describe the evolution of the Universe, starting with 10^{-4} seconds from the start of the Big Bang. At the moment $t = 10^{-4}$ seconds, the entropy of the observable part of the universe differs from the current entropy by no more than one order of magnitude.[33] One expects that at the initial stage of the Big Bang, the entropy of the world was very small. Such a possibility has been formulated by Penrose [51]. Thus, the sharp rise of entropy that defines the arrow of time appears to relate to the short times after the Big Bang, when the gravitational interaction becomes strong and when quantum effects are beginning to play a part — that is the quantization of space-time. This question has been examined by many, see for instance [51], but the result was disheartening — no satisfactory explanation was produced. If the growth of entropy that defines the arrow of time does indeed happen in the region of $\approx 10^{-33}$ cm, then — in accordance with what was said above — its cause will remain unknown.

In mathematics it is long since known as a fact that in some cases it is impossible in principle to determine the truth of a specific statement. There is a theorem — Gödel's theorem, proved in 1930 [52], which states that in any system of axioms in arithmetics a statement can be found in relation to which it is impossible to say whether it is true or false. In other words, in mathematics uncognizability has been strictly proven.

Until now I have been considering the question of the cognizability of the world from a position assuming (though not stating it in so many words) that civilization will exist for a fairly long time, say several thousands of years. But this is a very strong assumption; there are many reasons for civilization to perish much earlier: nuclear wars may start for some reason or another (a fanatic dictator comes to power in some country, terrorists acquire nuclear weapons, and so on), or some virus may appear in genetic science (unintentionally) and no means to withstand it can be found. Therefore it is possible that the current civilization will last another hundred or two hundred years. This may well explain the failure of all attempts to find extra-planetary civilizations: the factor $100/10^{13}$ (the time elapsed since the Big Bang) is rather short.

---

[33]The author is grateful to S.I. Blinnikov for this estimate.

5.3. Brief information about scientists mentioned in this book

- Aleksandrov, Anatoli Petrovich (1903–1994) — physicist, Academician, President of the USSR Academy of Sciences (1975–1986), Director of Institute for Physical Problems (1946–54) and Atomic Energy Institute (1960–1986), three times Hero of Socialist Labour.
- Artsimovich, Lev Andreevich (1909–1973) — physicist, Academician, co-author and close friend of the Alikhanov brothers.
- Berestetsky, Vladimir Borisovich (1913–1977) — physicist, Head of the Laboratory of Theoretical Physics ar ITEP (1966–1977).
- Bethe, Hans Albrecht (1906–2004) — theoretical physicist, Nobel Prize laureate, one of the leading participants of the American nuclear project.
- Bogoliubov, Nikolai Nikolaevich (1909–1992) — mathematician and theoretical physicist, Academician.
- Bjorken, James (1940–) — theoretical physicist, one of the founders of the quantum particle theory.
- Vinogradov, Aleksandr Pavlovich (1895–1975) — chemist, Academician, leader of chemistry in the Atom Project.
- Vannikov, Boris Lvovich (1887–1962) — head of the First Main Directorate at the Council of People's Commissars of the USSR (PGU), the organization in charge of producing atomic weapons, three times Hero of Socialist Labour.
- Vladimirsky, Vassili Vassilievich (1915–2002) — physicist, Corresponding Member of the Academy, Assistant Director of ITEP.
- Galanin, Aleksei Dmitrievich (1916–1999) — theoretical physicist, ITEP.
- Gell-Mann, Murray (1929–) — theoretical physicist, Nobel laureate, one of the founders of the quark theory.
- Gershtein, Semion Solomonovich (1929–2000) — theoretical physicist, Academician, Institute of High Energy Physics.

- Ginzburg, Vitali Lazarevich (1916–) — theoretical physicist, Academician, Head of the FIAN Theory Department, Nobel laureate.
- Glashow, Sheldon (1932–) — theoretical physicist, Nobel laureate, one of the founders of the electroweak theory.
- Gurevich, Isai Izrailevich (1912–1992) — physicist, Corresponding Member of the Academy. Fundamental work on nuclear physics, on theory of nuclear reactors.
- Yelian, Amo Sergeevich (1903–1965) — prominent Soviet figure in State and industry affairs, Head of many leading enterprises of the Soviet military-industrial complex.
- Ioffe, Abram Fedorovich (1880–1960) — physicist, Academician, Vice-President of the Academy of Sciences (1942–1945), Director of Leningrad Physical-Technical Institute, initiator of nuclear research in the USSR.
- Kantorovich, Leonid Vitalievich (1912–1986) — mathematician, economist, Nobel laureate.
- Kikoin, Isaak Konstantinovich (1921–1986) — physicist, Academician, Head of project on the separation of isotopes in the Soviet atom project.
- Kronrod, Aleksandr Semenovich (1921–1986) — mathematician, Head of the Mathematics Laboratory at ITEP.
- Lederman, Leon (1926–) — experimental physicist, Nobel laureate, Director of Fermilab (1979–1989).
- Low, Francis (1928–2014) — theoretical physicist, author of fundamental work on quantum electrodynamics.
- Markov, Moisei Aleksandrovich (1908–1994) — theoretical physicist, Academician.
- Marshak, Robert (1919–1992) — theoretical physicist, published works on weak interactions. Founder of international conferences on high energy particle physics.
- Nikitin, Sergei Iakovlevich (1916–1990) — experimental physicist at ITEP.
- Okun, Lev Borisovich (1929–2015) — theoretical physicist, Academician.

- Oppenheimer, Robert (1904–1967) — theoretical physicist, Head of the American Atom Project.
- Pauli, Wolfgang (1900–1960) — theoretical physicist, Nobel laureate, one of the founders of quantum theory.
- Pontecorvo, Bruno Maksimovich (1913–1993) — physicist, Academician, worked with E. Fermi. In 1950, moved illegally/defected to the Soviet Union.
- Rudik, Aleksei Petrovich (1921–1993) — theoretical physicist at ITEP, a close friend of the author's.
- Rumer, Yuri Borisovich (1901–1985) — theoretical physicist, co-author and close friend of L. Landau, arrested on the same day as Landau in 1938.
- Ter-Martirosian, Karen Avetovich (1922–2005) — theoretical physicist, Corresponding Member of the Academy, ITEP.
- Feinberg, Evgenii Lvovich (1912–2014) — theoretical physicist, Academician, FIAN.
- Feinberg, Saveli Moiseevich (1910–1973) — specialist in the physics of reactors, Atomic Energy Institute.
- Feynman, Richard (1918–1988) — theoretical physicist, Nobel laureate, one of the founders of modern quantum electrodynamics.
- Fermi, Enrico (1901–1954) — physicist, Nobel laureate, one of the leading physicists in the American Atom Project.
- Khalatnikov, Isaak Markovich (1919–) — theoretical physicist, Academician, co-worker of L. Landau, director of the Institute of Theoretical Physics named after Landau.
- Khariton, Yuli Borisovich (1904–1996) — experimental physicist, Academician, Head of science at Arzamas-16, three times Hero of Socialist Labour.
- Schwinger, Julian (1918–1994) — theoretical physicist, Nobel laureate, one of the founders of quantum electrodynamics.
- Shmushkevich, Ilia Mironovich (1912–1969) — physicist, Head of Laboratory of Theoretical Physics at the Leningrad Physical-Technical Institute, a close friend of I. Pomeranchuk.

5.4. Institutes mentioned in this book

- Institute of High Energy Physics — IHEP.
- Institute of Theoretical and Exeprimental Physics — ITEP, Laboratory No. 3, TTL.
- Kaiser Wilhelm Institute — KWI.
- Institute for Nuclear Physics — JINR.
- Institute for Transport Engineers — MEMIIT.
- Kapitsa Institute for Physical Problems — IPP.
- Kharkiv Institute of Physics and Technology — KIPT; formerly Ukrainian Institute of Physics and Technology — UPTI.
- Kurchatov Institute — Laboratory No. 2 Academy of Sciences; formerly Institute of Atomic Energy.
- Institute of Physics and Technology — IPT.
- Laboratory of Measuring Apparatus — LIPAN.
- Lebedev Institute — FIAN.
- Leningrad Institute of Nuclear Physics — LNPI (since 1992: B.P. Konstantinov Petersburg Nuclear Physics Institute — PNPI).
- Moscow Aviation Institute — MAI.
- Moscow Institute of Physics and Technology — MIPT.
- Moscow Higher Technical College (Bauman Institute).
- Polytechnic Institute — PTI.
- Radium Institute.
- Yerevan Physical Institute — YerPhI.

Bibliography

———————————— • ————————————

[1] *Istoria Diplomatii* [History of Diplomacy] Vol. 3, OGIZ, Moscow, 1945 (in Russian).

[2] L. Lopukhovsky, *Viazemskaya katastrofa 41 goda* [The Viazma disaster of 1941], "Yauza", "Eksmo", Moscow, 2006 (in Russian).

[3] V. Suvorov, *Icebreaker: Who Started the Second World War?* Viking Press/Hamish Hamilton, 1990.

[4] F. Soddy, *Atomic Transmutations*, New World, 1953.

[5] Andrei Sakharov, *Memoirs*, Knopf, N.Y., 1990.

[6] R. Rhodes, *The Making of the Atomic Bomb*, Simon and Schuster, N.Y., 1986.

[7] R. Rhodes, *Dark Sun: The Making of the Hydrogen Bomb*, Simon and Schuster, N.Y., 1996.

[8] A. Kruglov, *Kak sozdavalas' atomnaya promyshlennost' v SSSR* [How the Atomic Industry was Created in the USSR] ZNIIATOMINFORM, Moscow, 1995 (in Russian).

[9] *Istoriya sovetskovo atomnovo proekta* [History of the Soviet Atom Project], E.P. Velikhov editor, Moscow, 1997 (in Russian).

[10] E.L. Feinberg, *Epokha i lichnost'* [Epoch and Personality] Nauka, Moscow, 1999 (in Russian).

[11] V.L. Ginzburg, *O nauke, o sebe i o drugikh* [On Science, on Myself and on Others], Fizmatgiz, Moscow, 2003 (in Russian).

[12] D. Holloway, *Stalin and the Bomb: The Soviet Union and Atomic Energy, 1939–1956*, Yale University Press, 1994.

[13] J. Medawar and D. Pyke, *Hitler's Gift: Scientists who fled Nazi Germany*, Platens, London, 2001.

[14] P.A. Aleksandrov, *Akademik Anatoly Petrovich Aleksandrov: Priamaya rech'* [Academician Anatoly Petrovich Aleksandrov: Direct Speech], Nauka, Moscow, 2001 (in Russian).

[15] E. Amaldi, *The Adventurous Life of Friedrich Georg Houtermans, Physicist (1903–1966)*, SpringerBriefs in Physics, 2012.

[16] Ya.B. Zel'dovich and Yu.B. Khariton, *Delenie i tsepnoi raspad urana* [Fission and Chain Decay of Uranium], Uspekhi Fizicheskikh Nauk **23**, 329 (1940) (in Russian).

[17] I.I. Gurevich and I.Ya. Pomeranchuk, *Reaktorostroeniye i teoriya reaktorov* [Reactor Design and Reactor Theory] in *Proceedings of the First International Conference on the Peaceful Uses of Atomic Energy*, Geneva, 1955; USSR Academy of Sciences p. 220.

[18] A. Akhiezer and I. Pomeranchuk, *Vvedenie v teoriyu neitronnykh mul'tiplitsiruyushchikh sistem (reaktorov)* [Introduction to the Theory of Neutron Multiplication Systems (of Reactors)], IzdAt, 2002 (in Russian).

[19] A.D. Galanin, *Teoriya geterogennovo reaktora* [Theory of Heterogeneous Reactors], Atomizdat, 1971 (in Russian).

[20] A.D. Galanin, *Vvedenie v teoriu yadernykh reaktorov na teplovykh neitronakh* [Introduction to the Theory of Nuclear Reactors on Thermal Neutrons], Energoizdat, Moscow, 1990 (in Russian).

[21] A.D. Galanin and B.L. Ioffe, *Tiazholovodnyi reaktor-razmnozhitel' na tsykle U − Th*, [Heavy-Water Breeder Reactor on the $U - Th$ cycle] TTL Report No. 319, 1950 (in Russian).

[22] B.L. Ioffe and L.B. Okun, *O vygoranii goriuchevo v yadernykh reaktorakh* [On the Fuel Burn-up in Nuclear Reactors] Atomnaya energiya **1** No. 4, 80 (1956) (in Russian).

[23] *Sbornik pamiati akademika I.I. Alikhanova* [In Memoriam Academician I.I. Alikhanov], Nauka, Moscow, 1975 (in Russian).

[24] *Vospominaniya o L.D. Landau* [Remembering Academician L.D. Landau], I.P. Khalatnikov (editor), Nauka, Moscow, 1988 (in Russian).

[25] A. Livanova, *Landau* [Landau], Znanie, Moscow, 1983 (in Russian).

[26] K. Drobantseva-Landau, *Akademik Landau: kak my zhili* [Academician Landau: how we lived], Zakharov, ACT, 1999 (in Russian).

[27] G.E. Gorelik, *Moya antisovetskaya deyatel'nost' — Odin god iz zhizni Landau* [My anti-Soviet Activity — one Year in the Life of Landau], Priroda **11**, 93-104, 1991 (in Russian).

[28] S.F. Edwards, *A Nonperturbative Approach to Quantum Electrodynamics*, Phys. Rev. **90**, 284, 1953.

[29] B.L. Ioffe, L.B. Okun and A.P. Rudik, *The Problem of Parity Nonconservation in Weak Interactions*, JETP **5**, 328 (1957).

[30] L.D. Landau, *Conservation Laws in Weak Interactions*, JETP **5**, 336 (1957).

[31] J.H. Christenson, J.W. Cronin, V.L. Fitch and R. Turlay, *Evidence for the 2π Decay of the K_2^0 Meson*, Phys. Rev. Letters **13**, 138 (1964).

[32] B.L. Ioffe, *Limits of the Applicability of Weak-Interaction Theory*, JETP **11**, 1158 (1960).

[33] B.L. Ioffe, *Use of Current Algebra to Calculate Radiative Corrections to the β-Decay Constant in the Theory with Intermediate Boson*, Letters JETP **4**, 254 (1966).

[34] B.L. Ioffe and E.P. Shabalin, *Neutral Currents and the Limit of Validity of the Theory of Weak Interactions*, Sov. J. Nucl. Phys. **6**, 603 (1968).

[35] G. Gamov, *My World Line*, Viking Press, 1970.

[36] D.D. Ivanenko, *The Age of Gamov as Seen by a Contemporary*, in the Russian edition of [35], pp. 231–392.

[37] David C. Cassidy, *Uncertainty: the Life and Science of Werner Heisenberg*, N.Y., Freeman, 1992.

[38] N. Bohr to W. Heisenberg, *Draft of Letter*: www.nbi.dk/NBA/papers/docs

[39] R. Jungk, *Heller als tausend Sonnen*, Scherz, Bern, 1956.

[40] Nevill Mott and Rudolf Peierls, *Werner Heisenberg, 5 December 1901–1 February 1976*, Biogr. Mems. Fell. R. Soc. 1977, Vol. 23, pp. 212–251.

[41] H. Bethe, *The German Uranium Project*, Physics Today, July 2000.

[42] N. Ostroumov, *Armada, kotoraya ne vzletela* Voenno-istoricheskii zhurnal, 10, 1992, [The Armada that did not Take Off], Journal of the History of War, 10, 1992 (in Russian).

[43] A.D. Sakharov, *Violation of CP Invariance, C Asymmetry, and Baryon Asymmetry of the Universe*, Letters JETP **5**, 24 (1967).

[44] *Plasma Physics and Problems of Controlled Thermonuclear Reactions*, Moscow, ANUSSR Publishers 1954, Vol. 1, pp. 20–30.

[45] H. Feiveson and T. Taylor, Bull. Atom. Sci. **32**, 14 (1976).

[46] H. Bethe, in *Nuclear Power and its Full Cycle*, IAEA, Vienna, 1977, Vol. 7, p. 3.

[47] A. Radkovsky and A. Galperin, Nucl. Technol. **124** 215 (1988); A. Galperin, P. Reichert and A. Radkovsky, DCI Global Security **6**, 265 (1997).

[48] H. Ponomarev-Stepnoi, G. Lunin and A. Morozov, Atomic Energy, **85**, 263 (1998).

[49] B.L. Ioffe and B. Kochurov, *Preliminary Results of Calculations for Heavy-Water Nuclear-Power-Plant Reactors Employing* ^{235}U, ^{233}U, *and* ^{232}Th *as a Fuel and Meeting the Requirements of a Nonproliferation of Nuclear Weapons*, Physics of Atomic Nuclei **75**, 160 (2012).

[50] S. Weinberg, *The First Three Minutes*, Andre Deutsch, 1977.

[51] Roger Penrose, *The Emperor's New Mind*, Oxford University Press, 1989.

[52] K. Gödel, Monatshefte für Mathematik und Physik **38**, 173 (1931).

Printed in the United States
By Bookmasters